AF527857

Participatory Decision Support System for Horticulture Crops

Dileep Kumar
Officer-In-Charge
Regional Research Station - Jaisalmer
ICAR-CAZRI
Jaisalmer-345 001 (Rajasthan)

SSPH

SATISH SERIAL PUBLISHING HOUSE
403, Express Tower, Commercial Complex
Azadpur, Delhi-110033 (India)
Phone : 011-27672852, Fax : 91-11-27672046
e-mail : info@satishserial.com, hkjain1975@yahoo.com
Website : www.satishserial.com

Published by :

SATISH SERIAL PUBLISHING HOUSE

403, Express Tower, Commercial Complex, Azadpur, Delhi-110033 (INDIA)
Phone : 011-27672852 Fax : 91-11-27672046
E-mail : info@satishserial.com, hkjain1975@yahoo.com

© Publisher

ISBN 978-93-86200-71-6

© 2018. All rights reserved, no part of this publication may be reproduced, stored in a retrieval system or transmitted in any form or by any means, electronic, mechanical, photocopying, recording or otherwise without the prior written permission of the publisher and also the copyright rights of the printing, publishing , e-book of this edition and subsequent editions will vest with the publisher. All Computer floppies, CD's, e-book and in any other form relating to this book will be exclusive property of the publisher.

This book contains information obtained from authentic and highly regarded sources. Reasonable efforts have been made to publish reliable data and information, but the publisher cannot assume responsibility for the validity of all materials or the consequences of their use. The publisher have attempted to trace and acknowledge the copyright holders of all material reproduced in this publication and apologize to copyright holders if permission and acknowledgements to publish in this form have not been obtained. If any copyright material has not been acknowledged please write and let us know so that we may rectify it.

Composed, Designed & Printed in India

Preface

Participatory Decision Support System (PDSS) is an interactive, flexible, and adaptable computer based information system that utilizes decision rules, models, and model base coupled with a comprehensive database and the decision makers own insights, leading to specific, implementable decisions in solving problems that would not be amenable to management science models. Thus, a PDSS supports complex decision making and increases its effectiveness. The Skeleton of Information System Building an interactive computer programs that utilize analytical methods, such as decision analysis, optimization algorithms, programme scheduling routines, and so on, for developing models to help decision makers formulate alternatives, analyse their impacts, and interpret and select appropriate options for implementation.

Author

Contents

Chapter 1

Decision Support System – An Introduction

A **decision support system** (**DSS**) is a computer-based information system that supports business or organizational decision-making activities. DSSs serve the management, operations, and planning levels of an organization (usually mid and that may be rapidly changing and not easily specified in advance—i.e. Unstructured and Semi-Structured decision problems. Decision support systems can be either fully computerized, human-powered or a combination of both.

While academics have perceived DSS as a tool to support decision making process, DSS users see DSS as a tool to facilitate organizational processes. Some authors have extended the definition of DSS to include any system that might support decision making and some DSS include a decision-making software component; Sprague (1980) defines a properly termed DSS as follows:

1. DSS tends to be aimed at the less well structured, underspecified problem that upper level managers typically face;
2. DSS attempts to combine the use of models or analytic techniques with traditional data accessand retrieval functions;
3. DSS specifically focuses on features which make them easy to use by non-computer-proficient people in an interactive mode; and
4. DSS emphasizes flexibility and adaptability to accommodate changes in the environment and the decision making approach of the user.

DSSs include knowledge-based systems. A properly designed DSS is an interactive software-based system intended to help decision makers compile useful information from a combination of raw data, documents, and personal knowledge, or business models to identify and solve problems and make decisions.

Typical information that a decision support application might gather and present includes:

- Inventories of information assets (including legacy and relational data sources, cubes, data warehouses, and data marts),
- Comparative sales figures between one period and the next,
- Projected revenue figures based on product sales assumptions.

History

The concept of decision support has evolved mainly from the theoretical studies of organizational decision making done at the Carnegie Institute of Technology during the late 1950s and early 1960s, and the implementation work done in the 1960s. DSS became an area of research of its own in the middle of the 1970s, before gaining in intensity during the 1980s. In the middle and late 1980s, executive information systems (EIS), group decision support systems (GDSS), and organizational decision support systems (ODSS) evolved from the single user and model-oriented DSS.

According to Sol (1987) the definition and scope of DSS has been migrating over the years: in the 1970s DSS was described as "a computer-based system to aid decision making"; in the late 1970s the DSS movement started focusing on "interactive computer-based systems which help decision-makers utilize data bases and models to solve ill-structured problems"; in the 1980s DSS should provide systems "using suitable and available technology to improve effectiveness of managerial and professional activities", and towards the end of 1980s DSS faced a new challenge towards the design of intelligent workstations.

In 1987, Texas Instruments completed development of the Gate Assignment Display System (GADS) for United Airlines. This decision support system is credited with significantly reducing travel delays by aiding the management of ground operations at various airports, beginning with O'Hare International Airport in Chicago and Stapleton Airport in Denver Colorado. Beginning in about 1990, data warehousing and on-line analytical processing (OLAP) began broadening the realm of DSS. As the turn of the millennium approached, new Web-based analytical applications were introduced.

The advent of more and better reporting technologies has seen DSS start to emerge as a critical component of management design. Examples of this can be seen in the intense amount of discussion of DSS in the education environment.

DSS also have a weak connection to the user interface paradigm of hypertext. Both the University of Vermont PROMIS system (for medical decision making) and the Carnegie Mellon ZOG/KMS system (for military and business decision making) were decision support systems which also were major breakthroughs in user interface research. Furthermore, although hypertext researchers have generally been concerned with information overload, certain researchers, notably Douglas Engelbart, have been focused on decision makers in particular.

Taxonomies

Using the relationship with the user as the criterion, Haettenschwiler differentiates *passive, active,* and *cooperative DSS*. A *passive DSS* is a system that aids the process of decision making, but that cannot bring out explicit decision suggestions or solutions. An *active DSS* can bring out such decision suggestions or solutions. A *cooperative DSS* allows for an iterative process between human and system towards the achievement of a consolidated solution: the decision maker (or its advisor) can modify, complete, or refine the decision suggestions provided by the system, before sending them back to the system for validation, and likewise the system again improves, completes, and refines the suggestions of the decision maker and sends them back to them for validation.

Another taxonomy for DSS, according to the mode of assistance, has been created by Daniel Power: he differentiates *communication-driven DSS, data-driven DSS, document-driven DSS, knowledge-driven DSS,* and *model-driven DSS.*

- A communication-driven DSS enables cooperation, supporting more than one person working on a shared task; examples include integrated tools like Google Docs or Microsoft Groove.

- A data-driven DSS (or data-oriented DSS) emphasizes access to and manipulation of a time series of internal company data and, sometimes, external data.
- A document-driven DSS manages, retrieves, and manipulates unstructured information in a variety of electronic formats.
- A knowledge-driven DSS provides specialized problem-solving expertise stored as facts, rules, procedures, or in similar structures.
- A model-driven DSS emphasizes access to and manipulation of a statistical, financial, optimization, or simulation model. Model-driven DSS use data and parameters provided by users to assist decision makers in analyzing a situation; they are not necessarily data-intensive. Dicodess is an example of an open sourcemodel-driven DSS generator.

Using scope as the criterion, Power differentiates *enterprise-wide DSS* and *desktop DSS*. An *enterprise-wide DSS* is linked to large data warehouses and serves many managers in the company. A *desktop, single-user DSS* is a small system that runs on an individual manager's PC.

Components

Three fundamental components of a DSS architecture are:

1. The database (or knowledge base)
2. The model (i.e., the decision context and user criteria)
3. The user interface.

The users themselves are also important components of the architecture.

Development frameworks

Similarly to other systems, DSS systems require a structured approach. Such a framework includes people, technology, and the development approach.

The Early Framework of Decision Support System consists of four phases:

- Intelligence - Searching for conditions that call for decision;
- Design - Developing and analyzing possible alternative actions of solution;
- Choice - Selecting a course of action among those;
- Implementation - Adopting the selected course of action in decision situation.

DSS technology levels (of hardware and software) may include:

1. The actual application that will be used by the user. This is the part of the application that allows the decision maker to make decisions in a particular problem area. The user can act upon that particular problem.
2. Generator contains Hardware/software environment that allows people to easily develop specific DSS applications. This level makes use of case tools or systems such as Crystal, Analytica and iThink.
3. Tools include lower level hardware/software. DSS generators including special languages, function libraries and linking modules

An iterative developmental approach allows for the DSS to be changed and redesigned at various intervals. Once the system is designed, it will need to be tested and revised where necessary for the desired outcome.

Classification

There are several ways to classify DSS applications. Not every DSS fits neatly into one of the categories, but may be a mix of two or more architectures.

Holsapple and Whinston classify DSS into the following six frameworks: text-oriented DSS, database-oriented DSS, spreadsheet-oriented DSS, solver-oriented DSS, rule-oriented DSS, and compound DSS. A compound DSS is the most popular classification for a DSS; it is a hybrid system that includes two or more of the five basic structures.

The support given by DSS can be separated into three distinct, interrelated categories: Personal Support, Group Support, and Organizational Support.

DSS components may be classified as:

1. Inputs: Factors, numbers, and characteristics to analyze
2. User knowledge and expertise: Inputs requiring manual analysis by the user
3. Outputs: Transformed data from which DSS "decisions" are generated
4. Decisions: Results generated by the DSS based on user criteria

DSSs which perform selected cognitive decision-making functions and are based on artificial intelligence or intelligent agents technologies are called intelligent decision support systems (IDSS).

The nascent field of decision engineering treats the decision itself as an engineered object, and applies engineering principles such as design and quality assurance to an explicit representation of the elements that make up a decision.

Applications

One example is the clinical decision support system for medical diagnosis. There are four stages in the evolution of clinical decision support system (CDSS): the primitive version is standalone and does not support integration; the second generation supports integration with other medical systems; the third is standard-based, and the fourth is service model-based.

DSS is extensively used in business and management. Executive dashboard and other business performance software allow faster decision making, identification of negative trends, and better allocation of business resources. Due to DSS all the information from any organization is represented in the form of charts, graphs i.e. in a summarized way, which helps the management to take strategic decision. For example, one of the DSS applications is the management and development of complex anti-terrorism systems. Other examples include a bank loan officer verifying the credit of a loan applicant or an engineering firm that has bids on several projects and wants to know if they can be competitive with their costs.

A growing area of DSS application, concepts, principles, and techniques is in agricultural production, marketing for sustainable development. For example, the DSSAT4package, developed through financial support of USAID during the 80s and 90s, has allowed rapid assessment of several agricultural production systems around the world to facilitate decision-making at the farm and policy levels. Precision agriculture seeks to tailor decisions to particular portions of farm fields. There are, however, many constraints to the successful adoption on DSS in agriculture.

DSS are also prevalent in forest management where the long planning horizon and the spatial dimension of planning problems demands specific requirements. All aspects of Forest management, from log transportation, harvest scheduling to sustainability and ecosystem protection have been addressed by modern DSSs. In this context the consideration of single or multiple management objectives related to the provision of goods and services that traded or non-traded and often subject to resource constraints and decision problems. The Community of Practice of Forest Management Decision Support Systems provides a large repository on knowledge about the construction and use of forest Decision Support Systems.

A specific example concerns the Canadian National Railway system, which tests its equipment on a regular basis using a decision support system. A problem faced by any railroad is worn-out or defective rails, which can result in hundreds of derailments per year. Under a DSS, the Canadian National Railway system managed to decrease the incidence of derailments at the same time other companies were experiencing an increase.

❑❑❑

Chapter 2

Agri Crop Calender

Agriculture Crop Calendar for the Month of January

- Make half-moon terrace/tree basin for placement of manures and fertilizer in citrus
- Removal of water sprouts diseased and dried twigs cut and paste with Bordeaux paste.
- Harvesting of the turmeric and ginger completes.
- Main field preparation and transplanting of Boro paddy in low altitude region completes.
- Harvesting of Rabi maize in mid altitude and low altitude regions completes.
- Harvesting of soyabean in high altitude region completes.
- Land preparation for spring potato in high altitude region commences.
- T-Budding of citrus completes this is being done during feb-march.
- Harvesting of arecanut and black pepper continues.
- Weeding and jungle clearing of orchard completes.
- White washing, irrigation and manuring of fruit trees in the orchard continues.
- Pruning and grafting of fruit plants commences.
- Control of lichen and moss in the orchard with caustic soda.
- Watering of vegetable and fruit nurseries continues.
- Seedbed preparation for sowing of tomato in low altitude region commences.

- Land preparation and sowing of ladys finger, pumkin and french bean in low altitude region.
- Harvesting of mustard for seed (oil extraction) purpose in low altitude region commences.
- In plantation frame- forming prune in young tea commences, levelling and shifting completes.
- Collection of Paddy straw, well-rottedcow dung and soil for substrate and casing for button mushroom etc
- Planting of deciduous plants Apple,Plum ,Peach continuous.

Agricultural Activities for the Month of February

- First top dressing and irrigation of Boro paddy in low altitude region commences.
- Second top dressing and irrigation of wheat crop in low altitude region continues.
- Land preparation and sowing of spring potato completes by 2nd week in high altitude region.
- Land preparation and sowing of summer potato commences from 3rd week in high altitude region.
- Harvesting of arecanut and oranges completes.
- Pit digging and winter planting of fruit plant completes.
- White washing, irrigation and manuring of fruit trees in the orchard continues.
- Pruning and grafting of fruit plants completes.
- Watering of vegetable and fruit nurseries continues.
- Seedbed preparation for sowing of tomato in low altitude region completes.
- Transplanting of tomato seedlings in low altitude region commences.
- Land preparation for Kharif maize in low altitude and high altitude region commences.
- Harvesting of mustard seed (for oil extraction) in low altitude region completes.

- Land preparation and sowing of jute (capsularis) in low altitude region commences.
- Land preparation and sowing of cotton in low altitude region commences.
- Harvesting of winter potato in low altitude region completes.
- Harvesting of arecanut, black pepper, banana, carambola completes.
- Seed bed preparation for sowing of chillies, capsicum, tomato in high altitude region commences.
- Sowing of ladys finger, brinjal etc. in low altitude region completes.
- Top dressing and earthing up of ladys finger, pumpkin and French bean in low altitude region commences.
- Watering of seedbed and clonal nursery for tea commences.
- Seed bed preparation for spring propagation of tea completes.
- In plantation frame-forming prune complete, remove congestion, clearing of drains and fire line completes.
- Preparation of compost for button mushroom cultivation.

Agricultural Activities for the Month of March

- Second top dressing and irrigation of Boro paddy in low altitude region completes.
- Irrigation of wheat crop in low altitude region continues.
- Land preparation and sowing of summer potato completed by 2nd week in the high altitude region.
- White washing, irrigation and manuring of fruit trees in the orchard continues.
- Watering of vegetable and fruit nurseries continues.
- Transplanting of tomato seedlings in low altitude region completes.
- Weeding and earthing up of tomato seedlings (Feb planted) in low altitude region commences.

- Seed bed preparation and sowing of tomato in high altitude region commences.
- Land preparation and planting of Kharif maize in both high altitude and low altitude regions commences.
- Land preparation and sowing of Ahu paddy in low altitude region commences.
- Harvesting of Wheat in low altitude region commences.
- Land preparation and sowing of Jute (Capsularis) in low altitude region completes.
- Land preparation and sowing of Jute (Olitorious) in low altitude region commences.
- Land preparation and sowing of Mesta, Cotton in low altitude region commences.
- White washing, irrigation and manuring of fruit trees completes.
- Land preparation and sowing of chillies, capsicum, lettuce, onion, french bean, mustard, pumpkin in high altitude region commences.
- Top dressing and earthing up of ladys finger, cucurbits (Feb planting) in low altitude region.
- Construction of overhead shade for May June Tea propagation, filling of sleeves for planting of clones commences.
- Weed control and first application of fertilizer, tipping in mature tea plantations.
- Spawning of compost, incubation and preparation of casing (soil and cowdung) for button mushroom.

Agricultural Activities for the Month of April

- Final irrigation of Boro paddy and harvesting (December transplanted) in low altitude region commences.
- Harvesting of wheat in low altitude region completes.
- Top dressing and earthing up of spring and summer potato in high altitude region commences.
- Main field preparation and transplanting of tomato in low altitude completes.

- Weeding and earthing up of tomato (March transplanting) in low altitude region completes.
- Harvesting of tomato. Jan, Feb transplanting in low altitude region continues.
- Seedbed preparation and sowing of tomato in high altitude region continues.
- Main field preparation and transplanting of tomato in high altitude region commences.
- Land preparation and planting of kharif maize in low altitude, medium altitude and high altitude regions completes.
- Main field preparation and transplanting of Ahu paddy in low altitude region completes.
- Nursery bed preparation and sowing of Kharif paddy in medium and high altitude regions commences.
- Harvesting of seeds of cole crops, carrot, radish etc. for seed purpose in high altitude region commences.
- Land preparation and sowing of french bean, capsicum, onion, cucumber, pumpkin in high altitude continues.
- Land preparation and sowing of tumeric, ginger, in high and medium altitude regions completes.
- Land preparation and broadcasting of Kharif paddy in medium altitude and high altitude regions commences.
- Weeding and top dressing of jute (olitorious) in low altitude region commences.
- Land preparation and sowing of jute (capsularis) in high altitude region completes.
- Land preparation and sowing of mesta, cotton in low altitude region continues.
- Land preparation and sowing of soyabean in high altitude region commences.
- Layering of litchi, Assam lemon, guava etc. in medium altitude and low altitude regions commences.
- Planting of Assam lemon, Black pepper cuttings and banana suckers in low altitude region commences.

- Summer pruning of young trees commences and completes.
- Preparation of nurseries and sowing seeds of cole crops in high altitude region commences.
- Sowing of tea seeds in nursery completes.
- Apply Y.T.D. mixture in poor growth seedlings.
- Fertilizer application in both young and matured tea plantation completes.
- Harvesting of button mushroom.

Agricultural Activities for the Month of May

- Harvesting of Boro Paddy in low altitude region completes.
- Top dressing and earthing up of Kharif Maize (April planting) in high altitude, medium altitude and low altitude regions commences.
- Land preparation and sowing of Soya bean in high altitude region completes.
- Irrigation, top dressing of Jute (Capsularis) (April planting) in low altitude region commences.
- Land preparation and sowing of Jute (olitorious) in low altitude region completes.
- Land preparation and sowing of Mesta in low altitude region completes.
- Layering of litchi completes and inarching of mango commences.
- Land preparation and sowing of brinjal, chillies and capsicum in low altitude region completes.
- Fertilizer application and removal of mulching material in tea nurseries.
- Plucking of tea leaves (two leaves and a bud) in plantation commences.
- Plant protection measures against Mite and Aphids commences and completes.
- Cultivation of Oyster mushroom commences.

Agricultural Activities for the Month of June

- Harvesting of Ahu paddy in low altitude region commences.
- Harvesting of spring potato in high altitude region completes.
- Harvesting of Summer potato in high altitude region commences.
- Nursery bed preparation and sowing of tomato in high altitude region continues.
- Main field preparation and transplanting of tomato (May sowing) in high altitude region continues.
- Top dressing, earthing up and plant protection of tomato (May transplanting) in high altitude region commences.
- Top dressing and earthing of Kharif Maize in low altitude, medium Altitude and high altitude completes.
- Seed bed preparation and sowing of Sali paddy in low altitude region completes.
- Main field preparation and transplanting of Sali paddy in low altitude region completes.
- Nursery bed preparation and sowing of seeds for transplanted paddy in medium altitude and high altitude region completes.
- Transplanting of paddy in medium altitude and high altitude (April sowing) commences.
- Summer planting of fruit plants (temperate, sub-tropical and tropical fruits) Arecanut commences.
- Land preparation and sowing of Soya bean in high altitude region completes.
- Harvesting of French bean in high altitude region commences.
- Weeding, Top dressing and earthing up of jute, mesta and cotton in low altitude region continues.
- Top dressing and earthing up of ginger, turmeric in high altitude and medium altitude regions completes.
- Harvesting of jack fruit, mango, pineapple, lemon and guava in low altitude region commences.

- Harvesting of early varieties of apricot, plum and peach in high altitude region commences.
- Planting of rootstock in the nurseries for grafting commences.
- Top dressing and earthing of chillies, capsicum in high altitude region commences.
- Inarching of mango completes.
- Transplanting of brinjal, chillies in low altitude continues.
- Sorting of Tea seedling and Hardening of autumn clonal plant commences and completes.
- Planting of Tea and shade trees commences.
- Plucking of Tea leaves (two leaves and a bud).
- Cultivation of Oyster mushroom continues.

Agricultural Activities for the Month of July

- Harvesting of Ahu paddy in low altitude region completes.
- Harvesting of summer potato in high altitude region continues.
- Nursery bed preparations and sowing of Tomato in high altitude region completes.
- Main field preparation and transplanting of Tomato in high altitude region continues.
- Weeding plant protection top dressing and earthing up of tomato (June transplanting) in high altitude region commences.
- Harvesting of Tomato (April and May transplanting) commences.
- Harvesting of kharif maize in high altitude, medium altitude and low altitude regions commences.
- Main field preparation and transplanting of Sali paddy (June sowing) in low altitude region completes.
- Weeding and Top Dressing of Sali paddy (June transplanting) in low altitude region commences.
- Main field preparation and transplanting of paddy (June sowing) in high altitude and medium altitude regions completes.

- Main field preparation and transplanting of cole crops (cabbage, cauliflower, knol-khol) in high altitude region completes.
- Summer planting of Fruit plants(Temperate, Sub-tropical and Tropical Fruits)/Arecanut completes.
- Weeding and earthing up of Soyabean in high altitude region commences.
- Harvesting of French bean in high altitude region continues.
- Land preparation (sub soil mulching) of winter potato in high altitude region commences.
- Land preparation and sowing of Carrots, Radish and other root crops in high altitude region commences.
- Harvesting of Jackfruit, Temperarte fruits, Mango, Pineapple, Lemon continues.
- Transplanting of rootstock for grafting in the nursery completes.
- Weeding, top dressing and earthing up of Chillies, Brinjal, cole crops (June transplanted) commences.
- Preparation of lath flame, cellusing beds and planting of Tea cuttings (spring propagation) commences.
- Plucking of Tea leaves (Two leaves and a bud) continues.
- Land preparation and sowing of ladys finger, bitter gourd etc. in low altitude region commences.
- Land preparation and sowing of Lettuce, cabbage, cauliflower for seed production in high altitude region completes.
- Cultivation of Oyster mushroom continues.

Agricultural Activities for the Month of August

- Harvesting of summer potato (except for seed) in high altitude region completes.
- Main field preparation and transplanting of Tomato in high altitude region completes.
- Weeding, Top dressing, earthing up and plant protection for tomato in high altitude region continues.

- Harvesting of Tomato (May, June, transplanted) in high altitude region continues.
- Harvesting of Kharif Maize in high altitude, medium altitude and low altitude regions completes.
- Weeding and top dressing of Sali paddy (July transplanted) in low altitude region commences.
- Weeding and top dressing of Kharif paddy (July transplanting) in high altitude and medium altitude region commences.
- Harvesting of Kharif Maize in high altitude, medium altitude and low altitude region completes.
- Harvesting of French bean in high altitude region completes.
- Land preparation and sowing of winter potato in high altitude region commences.
- Land preparation and sowing of carrot, radish and other root crops in high altitude region completes.
- Harvesting of sub-tropical fruits like mango, pineapple, lemon, jackfruit continues.
- Land preparation and sowing of French bean in high altitude, medium altitude and low altitude regions commences.
- Weeding and earthing up of carrot, radish and other root crops in high altitude region commences.
- Weeding, top dressing and earthing up of chillies, brinjal, cole crops (June transplanted) in high altitude region completes.
- Land preparation and sowing of ladys finger, bitter gourd etc in low altitude region completes.
- Weeding, top dressing and earthing up of ladys fnger, bitter gourd etc (June planted) in low altitude region commences.
- Harvesting of apricot,plum,peach and other temperate fruits completes.
- Land preparation and sowing of rabi maize in low altitude and medium altitude regions commences.
- Land preparation and sowing of pea, mustard, garlic in high altitude region commences.

- Harvesting of ladies finger, cucurbits ,brinjal, chillies in low altitude region commences.
- Harvesting of Jute in low altitude region commences.
- Harvesting of pineapple,banana in low altitude region continues.
- Land preparation and sowing of carrot seed in high altitude region commences.
- Weeding and cleaning of nurseries, pruning of shade trees and panting tea cuttings (spring propagation) continues.
- Weeding (chemical and manual) and plucking of tea leaves (two leaves and a bud) in plantation continues.
- Cultivation of Oyster mushroom continues.

Agricultural Activities for the Month of September

- Weeding and top dressing of Sali paddy in low altitude region completes.
- Weeding and top dressing of Kharif Paddy in high altitude and medium altitude regions completes.
- Top dressing and earthing up of cole crops (July transplanted) in high altitude region completes.
- Harvesting of cabbage, knol-khol, chillies, brinjal, radish in high altitude region commences.
- Land preparation and sowing of winter potato in high altitude region continues.
- Weeding top dressing and earthing up of ladys finger, bitter gourd etc in low altitude region completes.
- Land preparation and sowing of rabi maize in low altitude and medium altitude completes.
- Land preparation and sowing of pea, mustard, garlic, lettuce, in high altitude region completes.
- Harvesting of ladys finger, cucurbits, brinjal, chillies in low altitude region continues.
- Harvesting of jute and mesta in low altitude region continues.

- Harvesting of pineapple, mango, jackfruit in low altitude region completes.
- Land preparation and sowing of rape and mustard in low altitude region commences.
- Land preparation and sowing of rabi maize in low altitude region completes.
- Harvesting of tomato in high altitude region continues.
- Land preparation and sowing of pea, mustard, garlic in high altitude region completes.
- Harvesting of chestnut, persimon commences.
- Harvesting of late varieties of pear (lagoon)in high altitude region completes.
- Harvesting of ginger, turmeric in high altitude and medium altitude regions commences.
- Planting of tea cuttings continues. Apply Y.T.D. mixture in tea nursery.
- Plucking of tea leaves (two leaves and a bud) continues, and apply 2nd dose of fertilizer in tea plantation.
- Cultivation of Oyster mushroom completes.

Agricultural Activities for the Month of October

- Harvesting of cabbage, knol-khol, radish, chillies, brinjal in high altitude region continues.
- Weeding, top dressing and earthing up of winter potato in high altitude region completes.
- Weeding, earthing up of root crops in high altitude region completes.
- Weeding, top dressing and earthing up of rabi maize in medium altitude and low altitude regions commences.
- Harvesting of ladys finger, cucurbits, brinjal, chillies in low altitude region continues.
- Harvesting of jute, mesta in low altitude region completes.

- Land preparation and sowing of rape and mustard in low altitude region completes.
- Weeding, top dressing and earthing up of pea, mustard, garlic, lettuce, in high altitude region commences.
- Harvesting of tomato in high altitude region completes.
- Harvesting of chestnut, persimon completes.
- Harvesting of cabbage knol-khol, chillies, radish, brinjal in high altitude region completes.
- Harvesting of cauliflower (early transplanted)in high altitude region commences.
- Harvesting of root crops (early planted) in high altitude region commences.
- Harvesting of ladys finger, bitter gourd in low altitude region (early planted) commences.
- Harvesting of Kharif paddy (early transplanted) in high altitude and medium altitude regions commences.
- Land preparation and sowing of lentil in low altitude region commences.
- Harvesting of ginger, turmeric in high altitude and medium altitude regions completes.
- Filling of soil in polythene sleeves and planting of tea cuttings, seed bed preparation for planting tea seeds commences.
- Pruning of clonal mother bushes for April-June propagation.
- Plucking of tea (two leaves and a bud) completes.
- Cultivation of Oyster mushroom completes. Preparation of compost for Button mushroom cultivation commences.

Agricultural Activities for the Month of November

- Nursery bed preparation and sowing of Boro paddy in low altitude commences.
- Harvesting of Sali paddy in low altitude region commences.
- Harvesting of Kharif paddy in high altitude and medium altitude completes.

- Top dressing and earthing up of rabi maize in medium altitude and low altitude region completes.
- Harvesting of cauliflower in high altitude continues.
- Land preparation and sowing of wheat in low altitude region commences.
- Harvesting of cabbage, knolkhol, radish, chillies, brinjal in high altitude region completes.
- Top dressing of rape and mustard in low altitude region commences.
- Harvesting of pea, mustard, lettuce in high altitude region commences.
- Harvesting of root crops in high altitude region continues.
- Harvesting of Ladys finger, bitter gourd in low altitude region commences.
- Harvesting of French bean in high altitude, medium altitude and low altitude region continues.
- Land preparation and second transplanting of cole crops, radish, carrot and leafy vegetables for seed purpose in high altitude region commences.
- Harvesting of Ladys finger, bitter gourd in low altitude region completes.
- Land preparation and sowing of lentil in low altitude region completes.
- Weeding and jungle cleaning of orchard plantation commences.
- Sowing of seeds of fruit plants for rootstock commences.
- Harvesting of mandarin orange/arcanut commences.
- Planting of tea cuttings in polythene sleeves completes by the middle of the month.
- Sowing of tea seeds in prepared nurseries commences.
- Light pruning and deep skiffing of tea bushes in the plantation commences.
- Harvesting of cotton in low altitude region commmences.

- Spawning of compost, incubation, preparation of casing (Soil and cowdung) for Button mushroom.

Agricultural Activities for the Month of December

- Nursery bed propagation and sowing of Boro paddy in low altitude Region completes.
- Main field propagation and transplanting of Boro paddy in low altitude region commences.
- Harvesting of Rabi Maize in medium altitude and low altitude regions commences.
- Land preparation and sowing of wheat in low altitude region completes.
- Weeding and top dressing of wheat (November planted) commences.
- Budding of citrus commences.
- Harvesting of oranges, banana, coconut/arecanut continues.
- Harvesting of winter potato in high altitude region completes.
- Harvesting of Sali paddy in low altitude region completes.
- Harvesting of cauliflower in high altitude region completes.
- Land preparation and sowing of wheat in low altitude region completes.
- Top dressing of rape and mustard in low altitude region completes.
- Harvesting of French bean, pea, mustard, lettuce in high altitude region continues.
- Harvesting of French bean, ladys finger, bitter gourd in medium altitude and low altitude regions completes.
- Harvesting of root crops in high altitude region completes.
- Land preparation and second transplanting of cole crops, radish, carrot and leafy vegetables for seed purpose in high altitude region.
- Weeding and jungle cleaning of orchard/plantations continues.

- White washing, irrigation and manuring of fruit trees in the orchard commences.
- Watering of vegetables and fruit nurseries commences.
- Budding of citrus commences.
- Sowing of tea seeds in nurseries continues.
- In tea plantation light pruning to be completed by middle of the month.
- Skiffing and levelling operations continues.

❑❑❑

Chapter 3

Agricultural Marketing

Agricultural Marketing in India

- Agricultural Marketing continues to be the mainstay of life for majority of the Indian population. It contributes around 25% of the GDP and employs 65% of the workforce in the country. The Government of India under the ministry of agriculture has also set up specific commodity Boards and export promotion council for monitoring and boosting the production, consumption, marketing and export of various agricultural commodities. Some of these organizations /boards are Cotton Corporation of India (CCI), Jute Corporation of India (JCI), Tea Board, Coffee Board, Spice Board, National Horticulture Board (NHB), National Agricultural Marketing Federation (NAFED), Agricultural Products Export Development Authority (APEDA), etc.
- Most agricultural commodity markets generally operate under the normal forces of demand and supply. However, with a view to protect farmer's interest and to encourage increased production, the government also fixes minimum support price (MSP) for some crops like Paddy, Wheat, Jute etc. as public policy instruments.
- The government also promotes organised marketing of agricultural commodities in the country through a network of regulated markets. There are 7161 Wholesale Regulated Markets in the country as on 31.3.2001. Besides, there are also 7293 Wholesale Markets and 27,294 Rural Periodic.
- The advent of regulated markets Regulated markets have helped in mitigating the market handicaps of producers/sellers at the wholesale assembling level. But, the rural periodic markets in general, and the tribal markets in particular, remained out of its developmental ambit.

- It was envisaged that physical markets with facilities and services would attract the farmers and the buyers creating competitive trade environment thereby offering best of the prices to the producers/sellers. Presently, about 23 Agricultural Produce Market Acts are in force in various States and UT in the country for enforcing and facilitating organized agricultural marketing functions.
- Besides in most States, there are a number of rural primary markets (Hats) which serves, on an average, an area within a radius of 8-16 KM. The secondary assembling market serves, on an average, about 500 Sq.Km.

Agricultural Marketing in North-East India

Distribution of Wholesale Assembling Primary and Regulated Markets in NER

Sl. No.	State	No. of Wholesale Markets	No. of Rural Periodic Markets	Total	No.of Principal Regulated Markets	No. of Sub-yard	Total
1	Assam	172	650	822	16	19	35
2	Arunachal Pradesh	-	50	50	-	-	-
3	Manipur	20	49	69APMC act not assed			
4	Meghalaya	101	82	183	2	-	2
5	Mizoram	8	35	43	-	-	-
6	Nagaland	16	80	96	-	-	-
7	Tripura	84	554	638	21	-	21
Total		401	1500	1901	39	19	58

Source: Directorate of Marketing and Inspection, Ministry of Agriculture, Govt. of India

The North Eastern region comprising of the above 7 (seven) States are characterised by hilly terrain, tribal population and shifting cultivation (jhum). Agriculture in the region is practised by predominantly small and marginal farmers who generally follow the traditional methods of agriculture resulting in low productivity. The situation is further aggravated by the lack of proper marketing and services, lack of adequate roads and transport facilities, lack of processing units, lack of organised marketing structures along with other factors have caused great inconveniences to the farmers who are compelled to dispose off their produce often at uneconomical and uncompetitive prices.

In most of the rural markets, facilities are either not available at all or not adequate. The primary markets in Assam, Meghalaya and Tripura do not even have basic infra structural facilities and are in fact mere meeting place for buyers and sellers. By and large, these markets are owned and managed by local bodies which rarely pay any attention to their development.

The important component of infrastructure necessary for growth of agricultural marketing are communication/ transport and storage facilities are utterly deficient. Almost 58% of the villages in the region do not have proper link roads. According to an estimate, 88% of the total arrivals in selected primary markets in Manipur, 70% in Meghalaya, 56% in Assam and 48% in Mizoram are carried on head load by the producers. The average distance covered varies from 5-10 KM. Road transport is the primary mode of transport from the producer's yard to the primary, secondary and terminal markets. Storage facilities in the rural areas and in the primary markets are either absent or insufficient. In Mizoram and Arunachal Pradesh, such facilities do not practically exist. In Assam, Manipur and Tripura, the storage facilities available are predominantly "katcha". Only in Meghalaya "pucca" storage facilities are available in large number of primary markets.

Post-harvest handling which are facilitating functions like sorting, grading and standardization, packaging, post-harvest treatments etc. are not commonly practiced. All these factors impact the shelf life in storage and long distant transportation. The losses in transit are estimated to be 35-51% in case of fruits and vegetables.

Market Regulation in Meghalaya

The State Agricultural Produce Marketing Act was enacted in the year 1980 and the State Agricultural Marketing Board was set up in 1983, to develop marketing infra structural facilities and to provide marketing support to the farmers in the State. In the year 1991, with the assistance of the Centre for Agricultural Marketing, Government of India located at Jaipur, detailed survey was conducted and a project profile for development of marketing infrastructure in Meghalaya was prepared. Accordingly, it was proposed to set up secondary markets in each District in the State which are called Wholesale Regulated Market. Thus, land was made available at Mawiong in the East Khasi-Hills District, Garobadha in the West Garo Hills District and Williamnagar in the East Garo.

□□□

Chapter 4

Agro-Entreprenuers of Meghalaya

Name of The Firm	Address / Contact	Business Activity
M/S. Queency Thangkhiew	Nongkseh, 3rd Mile Upper Shillong-793005 East Khasi Hills, Meghalaya	Floriculture
M/S. Biloris Warjri	Mawkriah-793121, East Khasi Hills, Meghalaya	Floriculture
M/S. Glaxoris Warjri	Mawkriah-793121, East Khasi Hills, Meghalaya	Floriculture
M/S. Tinolin	Mawkriah-793121, East Khasi Hills, Meghalaya	Floriculture
M/S. Morningstar War	Lawsohtun, Shillong-793004 East Khasi Hills, Meghalaya	Floriculture
M/S. Bernadette Khongsngi	Umran Diary,Ri-Bhoi District, Meghalaya	Floriculture
M/S/ Mansan Fruit Products	Industrial Area, Barapani, Meghalaya.	Fruit Processing
M/S. Regetta Food Pvt. Ltd.	Laitumkhrah, Shillong-793003 East Khasi Hills, Meghalaya.	Meat Food Products
M/S. Kara Industries	Pohkseh, Shillong-793003 East Khasi Hills, Meghalaya	Processed Products
M/S. Hunbait Industries	Jowai, Jaintia Hills, Meghalaya	Spices
M/S. J.G. Spice Pvt. Ltd.	Byrnihat,Ri-Bhoi District, Meghalaya	Ginger, Tumeric Processing
M/S. Nalari Tea Industries	Mawsyntai,Ri-Bhoi District, Meghalaya	Tea Processing

M/S. Areng Industries	Rongram,West Garo Hills District, Meghalaya	Tea Processing
M/S. Anderson Biotech Pvt. Ltd.	Lower Lachumiere Shillong-793001 East Khasi Hills, Meghalaya	Organic Manure Processing (Bio-Plus)
M/S. Merry Dew Wine	Upper Nongthymmai Shillong-793014 East Khasi Hills, Meghalaya Ph.: 9863117098	Wine
Mawphlang Distillery	Mawphlang, East Khasi Hills Meghalaya Ph.: 9863252192	Wine
M/S. Damad Food Processing	Pohkseh, Shillong, East Khasi Hills Meghalaya Ph.: 9863232138	Wine
ACE	Contact : 9436307419	Wine
Breezing Wine	Contact : 9856469145	Wine
M. Syiem (Forever Young)	Contact : 9863063490	Wine
M/s B. Kharshandi Nursery	Kyiem, Mylliem East Khasi Hills Ph.: 8575013446	Nursery for Temperate Fruits Plum, Pear, Peach etc.

❑❑❑

Chapter 5

Banana

Scientific Name *Musa* spp.

Common Name ***Khasi***: Kait; ***Assamese***: Kol; ***Bengali:*** Kola; ***Manipuri***: Laphoi

Family ***Musaceae***

Composition and Uses

- Banana is one of the delicious, cheapest and one of the best fruits of the world and the common man's fruit of this region.
- It yields highest amount of energy giving food per unit area.
- Soft ripe banana is usually the first solid food given to babies in this region of the country.
- Ripe banana is a rich source of carbohydrates, minerals like calcium, phosphorous, iron etc., and also contains appreciable amounts of vitamins, especially B and A.
- It also supplies some amount of proteins. Thus, it can be called a complete food.
- The common English term plantain is used for green banana which are grown only for vegetable purposes.
- Unripe banana as well as the inner core of the pseudo stem is commonly consumed as vegetables by the people of this region.
- The lower end of the inflorescence is also cooked as a vegetable while chopped banana stems are used as cattle feed. Its leaves are used as plates for serving meals in social functions in various parts of India including this region.
- Various products can be prepared from tables as well as plantain varieties.

- Banana fig is prepared by drying ripe fruits on domestic or large scale. Green banana or plantain is also dried and ground, and is commonly known as banana flour or plantain meal.
- It is very nutritious and easily digestible, and, therefore, can be given to children along with milk as a nutritious diet, because it supplies about 80 percent carbohydrate and 3.5 percent proteins per unit of dry weight.

Agro-climatic condition **Climate**

- Banana is basically a tropical plant but it has wide adoptability to climate ranging from wet tropical to dry subtropical.
- It can be grown successfully from sea level to an altitude of 1000 m in hills of the north eastern states, but varieties differ along altitude. Being a tropical crop, its water requirement is very high; it needs an annual rainfall of 2000 to 4000mm, evenly distributed over different parts of the year.
- In most parts of the north eastern hill region, banana is successfully grown as a rainfed crop.
- Drought is a limiting factor in banana cultivation. In banana growing areas, distinction has been made between hot season and cold season drought.
- In north eastern hill region, plants suffer considerably from winter drought.
- Cold in winter reduces vegetative growth of banana.
- Fewer leaves are produced and the rate of their emergence is considerably slowed down.
- Consequently, the crowns of the plants at the end of winter are much reduced and bunches are correspondingly slow to develop.
- This delay results also in the reduction of leaf area and thus the fruit may be exposed to severe scorching by the sun.

- In severe cold malformed bunches or incomplete fruits develop in subtropical latitudes.
- Periods of drought or low temperature reduces growth and flowering with the result that the peak production period is delayed.
- Another major climatic problem in hills is frost, which swiftly kills banana leaves and just one night with freezing temperatures is sufficient to ruin a plantation. Banana cannot be grown at all in areas regularly subjected to frost.
- The effects of occasional frost, however, can be circumvented by planting mainly on hill slopes rather than in the valleys. Secondly, if the frost is not very severe, it may affect a zone of air only a few feet above the ground, so in this case tall clones can escape frost.
- Thirdly, frost may be countered by the use of suitable heaters or fires for generating heat in the plantation or by sprinkling banana foliage with water having a temperature higher than that of air.
- Severe hail storms in higher altitudes sometimes do considerable damage to banana leaves.
- Wind also affects banana plantations.
- Light winds because a characteristic tearing of the leaves and severe winds either break or uproot banana pseudo stems, thus causing loss of fruits.

Soil

- Banana grows well in most soils except sandy, because in sandy soil, plants are easily uprooted.
- At the same time, moisture retention of sandy soil is also very poor.
- Virgin soils of forests are the best, because they can sustain a banana orchard for

several years without any extra supplement of nutrients. The best soils for its cultivation are the ones that are warm, well drained, moisture retentive and at least one meter deep.

Temperature
- The optimum temperature requirement varies from 20°C to 26°C but it can be produced satisfactorily within a temperature range of 15°-35°C without much damage to the crop.

Cultivars Chinichampa (neypoovan), champa (poovan), Jahaji (dwarf cvendish), borjahaji and grandenaine (giant Cavendish), Morthomon (sabri)

Propagation
- The banana plant has an underground modified stem, called rhizome which bears several buds.
- Each of these buds sprouts and forms its own pseudostem and a new secondary rhizome.
- These new daughter plants are known as suckers and are used after separation as vegetative sources for raising new plantations.
- Two types of suckers are generally found: *sword suckers* and *water suckers*. Sword suckers have pointed leaves and bear early while water suckers are broad leaved and bear late.
- Therefore, sword suckers of 3-4 months age and about 0.6 to 1.0 metre height with well-developed rhizomes are preferred for plantation.
- The suckers are separated from the mother plant before planting. Sucker size should be related to planting time and weather.
- Bigger suckers can be planted successfully during monsoon due to sufficient amounts of available moisture.
- In later part of rainy season, small suckers are not available. Bits of rhizomes may be planted in nursery for sprouting before transplanting, though plants in this case require little longer time to fruit. Recently, multiplication of virus free planting materials has been reported by tissue culture in vitro.

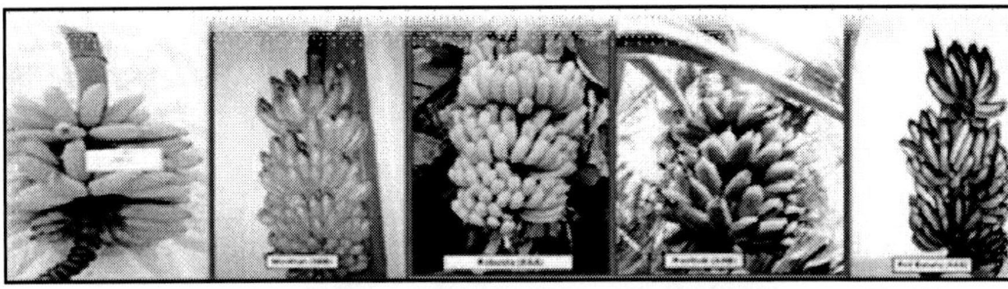

Co 1 Monthan Robusta Rasthali Red Banana

Karpooravalli Nendran Neypoovan Grand Naine

Planting

- Leveled or plain land is thoroughly prepared by ploughing before planting but on hill slopes, only individual pits are prepared for placing the planting material, instead of the whole area, to avoid soil erosion.
- When a new area is selected for plantation, cutting of forest plants and burning is practiced.
- Although it has been observed that banana grows remarkably well in burnt soil, it is not advisable for various reasons.
- On hill slopes pits 50 cm in size are prepared along the contour lines at a distance of 2.5-3.5 m. Spacing is affected by the growth habits of the plant and the fertility of soil.
- Spacing also depends on economic factors. For example, gross yield increases with increasing density, but grade declines slightly at a higher density, bunches decline very markedly in size and rate of maturity is delayed in ratoon crops.
- The time of pit preparation is linked with the time of planting. Generally, pits are dug 30-40 days before planting to allow the soil to get naturally sterilized.

- In case sufficient sunlight is not available, it is advisable to sterilize the open pits by burning fire wood. This practice also adds potassium in the pits.

- The pits are filled with 12-15 kg of FYM and 500-750 g of bone-meal along with top soil 8-10 days before planting, so that soil can settle properly in the pits.
- The usual period of planting is from April to June that is till the beginning of monsoon in north eastern region.

- Planting of banana can also be done in August-September. In late monsoon planting, growth of plants is checked by onset of the cold weather and fruiting is delayed.
- While planting, the suckers are placed in the pits in such a way that rhizomes remain in the pits at a depth of 25 to 30 cm, and the soil is firmly pressed.
- Planting of suckers is generally done during evening preferably on a rainy day.
- Before placing of suckers in pits, roots are pruned and in case of excessive vegetative growth, some leaves are also removed along with dry leaves.
- When the rhizomes or pieces of rhizomes are planted, care should be taken that soil should remain loose and moist for easy emergence of pseudo stems

Manuring

- Banana requires heavy manuring.
- There are three stages recognized in banana vizvegetative, floral and fruit development.
- Growth depends on preceding developments as well on contemporary events, and this leads to the view that an early development may limit later achievements.
- Thus vegetative growth in the first three months determines to a large extent the number and size of fruits ultimately borne by the shoot. Better vegetative growth i.e. a large leaf area is important fro a high yield.
- Proper growth and production of fruits can be achieved by judicious application of manures and fertilizers.
- When the plants put forth 3 to 4 leaves, they are given a compost of 7-8 kg of green leaves.
- In this region 250-300 g of nitrogen, 200-250 g of phosphorous and 750-1000 g of potassium should be given for successful production of banana fruits.

- It is better to apply half of the nitrogen in the form of organic manures and the remaining half in the form of chemical fertilizers, within three to five months of planting in two equal doses along with full amount of phosphorous and half in the form of chemical fertilizers, within three to five months of planting in two equal doses along with full amount of phosphorous and half of potassium.
- The rest of the potassium can be incorporated in the soil just before the emergence of flowers.
- In perennial banana, dry leaves, old banana pseudo stems and leaves are chopped and mixed in the soil to add organic matter.
- After Care: The banana rhizome produces many suckers continuously. These compete with the main plant and affect its growth and fruiting. They should, therefore, be removed in the vegetative stage.
- To ensure that the suckers do not sprout again, they are severed close to their rhizome without damaging it.
- Regular weeding is essential during the first 3 or 4 months. Some older leaves are allowed to remain hanging on the pseudostem to protect it from the sun.
- During rainy season, fast growing cover crops can be grown successfully.
- Intercropping with vegetables like cowpea, frenchbean, chilies, brinjal, spinach and cucurbits can give additional income to the growers without significant decreases in the yield of banana.
- Earthing is essential for individual plants after every heavy shower, and soil should be placed on the pseudostem.

- Weeding is essential for proper utilization of nutrients as well as fro effective control of pests and diseases.
- Diuron at higher dose like 2-3 kg/ha before emergence is effective in controlling weeds.

Cropping

- Banana comes to bearing stage 10-15 months after planting, depending upon the size and age of the plant, variety, time of planting as well as fertility status of the soil.
- Inflorescences of banana generally emerge at 12-14 leaf stage. Growth of fingers is slow in winter months while it is faster in summer.
- It takes about another 3 to 4 months for the bunches to be matured and ready for harvesting.
- When flower stalk starts emerging on the main plant, only the first sucker is allowed to grow, and the second suckers are allowed only just before harvesting, during the later phase of growth.
- Beyond these two suckers, any other suckers are not allowed, as they affect growth and production.
- The removal of unwanted suckers is known as desuckering.
- It is reported that spraying of GA3 on 35 or 55 days old banana bunches increases the size and weight of fingers.
- Spraying of 2, 4, 5-T orAncymidolorRacuza at 100 ppm proved more effective than GA3 or Ethrel in increasing finger size and bunch weight.
- The developing bunches on the plant should be shaded with leaves to protect them from sunburn.
- If bunches are too heavy, the plant should be supported with bamboo props.

- The fruit is harvested after the fingers begin to turn light green and the angular ridges on their surface start becoming roundish.
- The top leaves at this stage start drying up and flowers at the end of fruits can gradually be harvested from the plant itself, but for commercial purposes the entire bunch is usually harvested, leaving the stock intact for about 30 cm for easy handling.
- They are usually wrapped with banana leaves. The average yield in this region is about 9.33 tonnes/ha. But in well maintained conditions it can give a yield of upto 50-60 tonnes/ha.
- If temperature is not regulated inside the storehouses, banana ripens very slowly and irregularly.
- The basal end of the bunch changes colour and ripen first, followed by the distal end.
- A common practice to ripen banana is to smoke the fruits inside closed chambers.
- Banana bunches are arranged in a heap in a room which is filled to three fourth of its capacity.
- The heap is covered with dry banana leaves. Fires are lighted in the room for smoking.
- Smoking lasts about 18 to 24 hours in summer and about 48 hour in winter.
- On opening the chamber, the banana bunches are taken out and heaped again in well ventilated godowns with roofs, where they change their colour within 2-3 days.

- Artificial ripening with a better colour development can be achieved by spraying Ethrel at a concentration of 1000 ppm on mature fruits. When spraying is done inside gas chambers, the concentration is reduced to 100 ppm.
- A post harvest dipping of banana fruits in an IAA solution also hastens ripening. Use of Calcium Carbide (CaC) (10-15 g) in paper lined baskets shorten the ripening time.

IPM/ Diseases Diseases and Pests

Bunchy top: It is a very serious disease of banana in the north eastern region, especially in Jahaji and Champa varieties.

- This disease is caused by a virus which is transmitted by an aphid (*Pentalonianig-ronervosa).*
- In the affected plants, leaves become short and bunch together at the top of the pseudostem to form a rosette.
- Growth is stunted and fruit production fails. Once a plant is attacked by this virus, it cannot be cured.
- However, other plants can be prevented from having it by removing and destroying the infected plants along with their rhizomes and spraying systemic insecticides like metasystox or Dimecron

at the rate of 0.1-0.5 percent on the healthy plants to kill insect vector.

- Strict quarantine regulations should be followed in disease free areas.

Panama disease (Banana wilt):

- This is a soil borne fungal disease caused by *Fusarium oxysporum F. cubense.*

- Progressive browning and falling of leaves and apance of black streaks on underground stems are the characteristic symptoms.
- Control measure is to remove infected plant parts, and application of 0.5-1 kg of lime in each pit.
- Soil drenching with fungicide vapour (850g in 100 litre of water) and mercuric chloride (3000 ppm) or 1 percent solution of formathion gives effective results.
- Highly infected soils should not be replanted with banana at least for 3-4 years.

Leaf spot: It is caused by the fungus Mycosphaerellamusicola.

- Light yellow to brown leaf spots occur, usually starting from margins and spreading inwards parallel to the veins.
- Infection occurs through stomata of young leaves. It can be controlled by spraying with (0.2%) blitox, dorsal, falcidin and Bavistin.

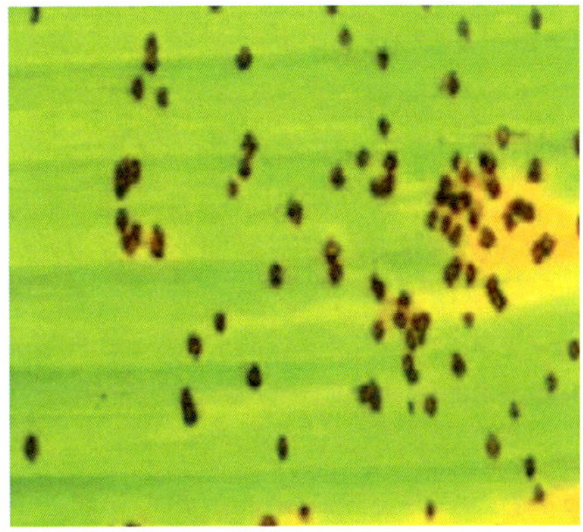

- Two percent linseed oil and Bordeaux mixture has been found to reduce the incidence of this disease.

Pseudostem heart-rot:.

It is also fungal disease caused by *Botryodiplodia, Gloeosporium* and *Fusarium* spp.

- The disease is characterized by the presence of heart leaves with part of the lamina missing or decayed.
- In severe cases, the inner leaves of the crown first turn yellow, then brown and finally die.
- To prevent spread of disease, spraying of captan or Dithane M-45 or Dithane Z-78 is effective.

Main stalk rot:tgA black soft rot begins at the proximal end of the main stalk of the bunch. Affected parts shrivel and dry up.

- Premature ripening of fruits often occurs when stalk rot is severe. It can be controlled by spraying of Dithane M-45 or Dithane Z-78.

Pitting disease (Pyricularia-grisea):

- Round sunken pits ap on the fruit as it approaches maturity or may also ap after harvesting.

- The centre of the pit may split the fruit skin. It can be easily prevented by regular weekly spraying of Dithane M-45.

Brown spot (Cercosporahayi):

- Brown spots occur on the rachis and fingers.

- The spots are pale to dark brown with irregular margins surrounded by water soaked tissue. The control measure is same as that of pitting disease.

Stem borer (Odoiporus longicollis): This is a very serious insect-pest which bores through the sheaths, eats into the pseudostem and also into the rhizome causing considerable damage in certain varieties like Champa, Malbhog, and sometimes Jahaji, in this region. Treatment with celphos at the rate of 3 tablets per plant can effectively control this pest. The larvae of *Cosmopolites sordidus* also bore into the underground rhizomes and weaken the plants.

Fruit and leaf beetle (Colaspishy pochlora):

- It feeds on the epidermis of leaves, skin of young fruits and inflorescence, causing corky tissues to develop on the surface of the fruits and leaves.

- Severe scarring of fruit skins leads to the production of underdeveloped fruits which fetch fewer prices in the market. Malathion (0.05%) sprays just before and after opening of the inflorescence can control this pest.

Banana aphid (Pentaloniani-gronervosa):

- This insect is particularly important, as it is the vector of virus causing bunchy top disease.
- It can be controlled by spraying rogar, Malathion and metasystox.

Nematode (Radopholussimilis):

- It is widely distributed in the banana growing areas.
- Affected plants do not respond to manuring and cultural operations.

- Before planting, suckers should be treated with Furadan granular nematicide. The nematicide is placed around the plants and is carried down into the soil by rain.

❑❑❑

Chapter 6

Broccoli

Introduction

- The vegetable 'broccoli' is a member of the Cole Crop group and is classified as a botanical variety of the species *Brassica oleracea*.
- Readers of horticultural trade literature published in the United Kingdom should be aware that the crop in that country is referred to as 'calabrese'.
- The term 'broccoli' in the U.K. is often applied to a winter-hardy cauliflower, planted in the field in the late summer and maturing to a normal white curd the following spring.

Varieties Solan Green Head, KTS, Lucky, Fiesta, Pushpa, Aiswarya

Climate It is a cool season crop but sensitive to very low and high temperature. The head becomes loose in high temperature.

Temperature Temperature range of 10-25°C is optimum for its proper growth, while 15-20°C for heading stage.

Soil Broccoli grows best on a well-drained, medium to heavy soil with high organic matter content. Broccoli requires moist soil for fast and proper growth. The shoots become more fibrous under dry soil. It does well in pH range of 5.0 to 6.5.

Nursery raising

- Plough the nursery beds thoroughly and mix with well rotten FYM
- Treat the seeds with Thiram or Captan @ 2.5g/kg of seed before sowing.

- Sow the seeds in the beds during the month of September at a spacing of 8-10 cm between lines, 2-3 cm between seeds and 1-1.5 cm deep.
- After sowing irrigate lightly and protect from heavy rains.
- Keep the beds free from weeds.
- About 400-500g seed is sufficient to raise seedlings for one hectare.

Transplanting The seedlings become ready for transplanting after 4-6 weeks of sowing. The planting of over mature seedlings should be avoided. Seedlings are transplanted 45cm apart within and between the rows. In very rich soils, spacing can be reduced to 45cm X 30cm to avoid stem hollowness due to rapid plant growth.

Manures and Fertilizers Apply 15-20 tonnes FYM, 80 – 100kg N, 80kg P and 60kg K. Full dose of P, K and half of N are applied at the time of land preparation. The remaining half dose of N should be top dressed in 2 equal split doses. The first is applied 4-5 weeks after transplanting, whereas the second dose before head formation.

Irrigation Broccoli needs sufficient moisture in the soil for uniform and continuous growth of plants. Therefore, frequent irrigation at 10-15 days is given depending upon weather conditions. The dry conditions adversely affect the quality and yield of shoots by being more fibrous. On the other hand, water logging condition depresses plant growth.

Intercultural Operations The crop should be kept free of weed. Shallow hoeing should be done to remove weeds and also for breaking the surface crust to facilitate better aeration and water absorption. Since it is a shallow-rooted crop, hoeing should not be done beyond the depth of 5-6cm to avoid injuries to the roots. A light earthing-up at final hoeing is beneficial.

Harvesting The heads having 10-15cm stems should be harvested with a sharp knife when its bud cluster is green and compact. If harvesting is delayed the bud cluster becomes loose. Pick regularly to ensure the heads are at their tastiest and to encourage the formation of side-shoots.

Yield Average yields of 100-150 Q/ha can be obtained depending upon the variety.

Pest

1. **Aphids**

   These are greenfly and blackfly which will be clearly visible on the leaves and especially tender young shoots. These seriously weaken the plant.

   Control:

   Protect the crop with Garlic fire spray preparation

2. **Caterpillars**

   These are visible as they much their way through the leaves. The cure is simply to pick them off by hand.

3. **Cabbage Gall Weevil**

   Stunted growth and misshapen roots with maggots appearing inside swellings near roots.

   Control:

   - Remove and destroy infected plants if serious.
   - If attack is only minor then harvest affected plants first.

4. **Cabbage Diamond-black Moth**

   Inner leaves get riddled with caterpillars showing round transparent cuticular patches caused by feeding.

   Control:

   a) Spray Bt product like Delfin 3G @ 1 gm/litre.

   b) Spraying of neem based formulations @ 4 ml/litre give good result.

Diseases

1. Clubroot Leaves turn bluish and wilting turning into a weak dying plant. When inspected, the roots will appear to be thick and distorted mass.

   Control

   Improve crop rotation and never grow broccoli in this area again for few years.

2. **Downy Mildew:** Light grey powdery patches on the leaves and shoots, normally appearing in spring. The mildew will get right into the plant, eventually killing it.

Control

i) Neem seed kernel 5 ml/litre spray after 25-30 days after transplanting.

ii) Neem oil 5 ml/litre -15 days after 1st spray of Neem seed kernel.

iii) Neem oil 5 ml/litre - 30 days after Neem seed kernel spray. Physiological Disorders:

1. **Whiptail**

Deficiency of molybdenum cause whiptail in which the lamina of the newly-formed leaves become leathery, irregular and consisting of only the mid-rib

Control

a) This can be corrected by soil application of 1-1.5 kg molybdenum before planting.

b) Foliar spray of 0.01% solution of Ammonium molybdate helps to check this disorder.

2. **Browning Head**

It results from boron deficiency. First water-soaked areas appear on bud clusters which in turn pinkish or rusty-brown in advanced stages resulting in rotting.

Control

a) Soil application of Borax or Sodium Borate @ 20 kg/ha will Prevent the disorder.

b) Foliar spray of 0.25-0.5% solution of borax is more effective than the former one, especially when the deficiency is acute. The affected portion does not recover fully but helps in appearance of new, healthy bud clusters.

❑❑❑

Chapter 7

Chilli or Capsicum

Botanical Name

1. *Capsicum Annum* L. (Fruits are long or small, but pungency is high).
2. *Capsicum frutescence* L. (Large and inflated variety with very little pungency).

Family Solanaceous

Chilli is one of the most valuable crops of India. The crop is grown largely for its fruits all over the India. It is used in India as a principle ingredient of various curries, and chutneys. It is also used for vegetables, spices, condiments, sauces and pickles. Dry chillies are used for curry powder. Red colour in chili is due to "Capsanthin". Pungency in chillies is due to the active constituent "Capsaicin", an alkaloid, is extracted from chillies and is used to medicine

Climate

The chili is a plant of tropical and sub-tropical region -It grows well in warm and humid climate and a temperature of 20°C to 25°C. Low moisture in soil during blossom development and fruit formation causes the bud, deblossom and fruit drops. Excessive rainfall is detrimental to the crops, because it brings about defoliation and rotting of the plant. As a rained crop, it is grown in areas receiving an annual precipitation of 25-30 inches.

Soil and its Preparation

Chilli can be grown in all type of soft but the sandy - loam, clay loam and loam soils are best suited for it, the soil must be well drained and well aerated. Acidic soils are not suitable for chili cultivation.

The land is prepared by giving 2-3 ploughings and clod crushing after each ploughing. Compost or FYM @ 150-200 quintals should be spread and mixed well in the soil at least 15-20 days before sowing. At the last ploughing 0. H. C. @ 8-10 kg per acre of Aldrin or Heftaf @ 10-15 kg per acre should be applied to the soil to protect the crop from white ants and other soil pests.

Sowing Time and Seed Rate

In case of chilli crop, for kharif it sown in May-June and for summer crop, it is sown in the month of January. I to I 1/2 kg seed is required for 1 hectare area.

Layout & Spacing

Ridges and furrow type of .layout is used. Seedlings are raised on raised bed. Spacing for rained crops is 60 x 45 cm & for irrigated crops is 60 x 60 cm.

Manures & Fertilizers

9 to 10 tones / hectare of FYM or compost is applied at the time of field preparation. In chilli for rained crop 50 kg N and 25 kg P should be applied 1/2 dose of N full dose of P applied at the time of transplanting. Remaining Yi dose of N applied 30 days after transplanting for .irrigated crop 100 kg N, 50kg P & 50 kg K should be applied per hectare. Fertilizers are applied in four equal doses. First applied at the time transplanting remaining doses are applied at 4th, 111 & 13' week after transplanting.

Improved varieties: N.P. 46, JwaJa, G-3, CA-960, Pant C-t X - 235, AKC - 79-18, Parbhani Tejas. Local: Achalpuri, Bhavapuri, Vashirn, Malkapuri.

Chilli (*Capsicum annuum*)

1. Agnirekha

   It is a derivative of the cross Dondaicha x Iwala Released for green fruits by MPKV, Rahuri in 1992, for summer season. Further it is also recommended for cultivation by Project Directorate of Vegetable Research in 1996 for Zone VII. Plants tall and spreading, fruits long, bold, smooth and light green in colour.

Duration is 180 days. Average yield of green" fruit is 250 quintals per hectare. Average yield of dry chilli is 27 quintals per hectare. (MPKV, Rahuri)

2. Musahvadi

Developed by selection from local collection of village Musahvadi, Dist. Ahmednagar (M.S.). It is released in 1988 by MPKV, Rahuri and also by the Central Sub Committee on Crop Standards, Notification and Varietal Release for Zones IV, V, VII and VIII in 1987 at the national level. Plants are tall, spreading habit with dark green foliage. Fruits are smooth and medium long. Mature fruits dark green with black patches, ripe fruits dark red with retentive colour. Duration is 180 days. It is tolerant to die back and wilt diseases. Average yield of red dry chilli fruits is 19 quintals per hectare. (MPKV, Rahuri)

3. Phule Jyoti

It is released in 1995 for kharif season. Plants are tall with spreading and branching habit from ground level. Leaves broad and dark green in colour. Fruits are smooth, medium long, borne in cluster of 5-6 with pendent habit. It is tolerant to thrips, mites and wilt. Average yield of green fruits 250 quintals per hectare and that of red dry fruits is 24 quintals per hectare. (MPKV, Rahuri)

4. Pbiile Sai

Developed by selection from the cross Pani C-l x Kamandalow. It has high yield potential under rain fed conditions. It retains the colour in storage for 6-7 months. It is moderately resistant to thrips and anthracnose. (MPKV Rahuri)

5. Suryamukhi

It is released in 1996. Plants tall, spreading and branching habit from ground level, leaves broad and dark green. Fruits smooth, medium long, borne in cluster of 5-6 and upright erect. It is tolerant to thrips and wilt. Average yield of green fruits is 190 quintals per hectare and yield of red dry fruits is 22 quintals per hectare. (MPKV, Rahuri)

6. Phule Mukta

It is developed by employing pure line selection and released by MPKV Rahuri. It has dark green foliage, plants tall and medium spreading. Fruits are small, dark green and smooth. Ripe fruits are dark red in colour. Duration is 210 days. Resistant to powdery mildew and Fusarium wilt. It is tolerant to leaf curl, thrips and mites and resistant to lodging, suitable for both the kharif and summer seasons. Average yield of dry chilli is 23 quintals per hectare. (MPKV, Rahuri)

7. Surkta

   Surkta has been released for Vidarbha region both for the rain fed conditions of kharif season. It produces bright deep red fruits. Fruits are very hot in taste. It gives about 35 per cent higher fruit yield compared to variety CA-960. Yield of red ripe fruits is 68 quintals per hectare. (PDKV, Akola)

8. Jayanti

   It is notified during the year 1996-97. It has been released for Vidarbha and at the national level under irrigated conditions. Plants are medium to tall in height with light green foliage. Fruits are medium to long in length (9-10 cm) with notch on 1/3rf portion from the tip of the frill. Fruits are pale green, white – green and turn red on ripening. Yield of red ripe chilli is 18-20 quintals per hectare. (PDKV, Akola)

9. Parbhani Tejas

   It is released in 1992 for Marathwada region for rain fed as well as irrigated conditions for green as well as red dry fruit production. It has pungency, fruits are longer (maximum 22 cm). Yield of dry red fruit is 18 quintals per hectare. Average yield on 20 farmers' field has been found 11.56 quintals per hectare as compared to 7.63 in case of NP-46. (MAU, Parbhani)

10. Kankan Kirti

    Plants are dwarf with 50-60 cm height and dark green leaves. Dark green fruits turn red when mature, mild pungent. It yields 12-14 tones per hectare of green chilli. Fruits are dark green, lustrous, good keeping quality and good export potential. (BSKKV, Dapoli)

Irrigation

Chilli is grown both as rain fed and irrigated crop. First irrigation is given after the transplanting and subsequent irrigations are given

5-7 days interval depending on weather and conditions of soil during summer and rainy season and after. Every 10 to 15 days in winter. The maintenance of uniform soil moisture is essential to prevent blossom .and fruit drops. Bosaocn and Garibaldi (1971) reported that the irrigation when soil moisture tension exceeded one atmosphere increased the yield of the unshared plant, but the number of fruits was unaffected. Flower drop in chillies is a great problem and it depends on the high temperature, low moisture availability, shading and light intensity. Flowers and/or flower bud abscission was increased under short day (day length 12 hours) and high temperature. Spraying the chilli with Planofix (NAA) at 10 ppm at flower initiation stage and 15 days later (peak flowering) reduces the flower drop and increases the fruit set.

Interculture Operations

2-3 shallow hoeing should be given to the soil to kill the weeds and provide soil mulch during early stages of growth- Application of weedicides for controlling the weeds is found effective. Lasso @ 1.5 litre per hectare with one hand weeding or Tok-E 25@ 2 litres per hectare with one hand weeding were effective in controlling the weeds.

Harvesting

1. Chillies which are used for vegetable purposes are generally harvested while they are still green but full grown.
2. Chillies are harvested at red stage for caning purpose. Chillies used for drying are harvested at full ripe-stage.

Yield

The yield varies according to the system of cultivation. The yield of dry chillies of rain fed crop is 200-400 kg and that of irrigated crop is 600-1000 kg per acre. The proportion of dry to fresh japed chillies varies from 25-40 per cent.

Insect Pests of Chilli

1. Chilli Thrips

   Thrips is the common pest which effects the crop throughout its life cycle. But they are more severe when plants begin to flower.

These small insects suck the sap from the foliage and lacerate the leaf tissue, which result in curling of leaves and fall down of flowers prematurely.

Control Measures

Thrips can effectively be controlled by spraying carbaryl 50 W @ 3 gm or Zolene @ 3 ml or Dimethoate (Rogor 30 EC) or Monocrotophos (Monocil) tgi 1 ml per litre of water at fortnightly interval

2. Pod Borer

The caterpillar oat leaves and later on bores the pod, which result in the deterioration of quality and market price of the product.

Control Measures

The control measures are timely spraying the crop with quinalphos (Exalux-25 EC) @ 4 ml or carbaryl 50 W, (Sevin 50 W) @ 3 gm per litre of water, starting from flower bud formation.

3. Aphids

Aphid suck the dap from the plants; they generally attack the crop in winter months and at the later stages of the crop. The quality of the produce is spoiled by imparting blackish colour to the Calyx and pods. They also serve as a Vector to Virus,

Control Measures

The aphids can effectively be controlled by spraying the crop with Dimethoate (Rogor-30 EC or Methyl Parathion (Metacid 50EC) @ lm (or Phosphamidon (Demicron-100 EC) @ 0.5 ml \per litre of whiter.

Diseases of Chilli

1. Damping Off

It is a serious disease of brinjal seedlings and mainly occurs in nursery bed. The disease infected seedlings rot at ground level and then the plants fall over ground. The seedlings die in patches.

Control Measures

1. The seed bed should be treated with Formalin before sowing of seeds.

2. The seeds should be treated without water (30 minutes at 52° C) or Cerasan or Agrosan G.N. before sowing of seed.
3. The seedlings in the nursery should be sprayed with any fungicides at a regular interval.

2. Bacterial Leaf Spot Small dark, greasy spots are formed on leaf, petiole and tender parts, of the plant. Water soaked spots appear on green fruits. In severe cases the leaf may drop off arid cause considerable loss to the crop.

   Control Measures

   Spraying Agrimycin - 100 k at 200 ppm plus copper oxychloride 0.3 per cent controls the disease effectively.

3. Anthracnose

   Dark sunken spots are formed of fruits and pink or dark coloured dots appear in the centre of the sunken spots. Due to this spots, the fruits rot and fall. The fungus may cause "Die back" of the twigs also. Die back disease attacks mainly the upper portion of the plants spreading gradually from the top to downward; as a result the branches dry up. Moist weather, shade and heavy dew favour the occurrence of the disease.

   Control Measures

   The control measures are the treatment of seed with Cerasan before sowing, removing and burning of attacked plants or branches and spraying the disease affected crops with Mancozeb (Dithane M-45) @ 2.5 gm per litre of water.

4. Leaf Curl

   The disease affected leaves becomes small in size accompanied by downward curling. The leaves may fall off in case of sever attack. The disease usually spreads through insect vectors such as thrips and aphids etc.

   Control Measures

   Control of insect vectors by spraying the crop with Dimethoate (Rogor -30 EC) or Monocrotophos (Monocil) @ 1 ml per litre of water indirectly helps to check the spread of this disease.

❑❑❑

Chapter 8

Cucumber

The cucumber responds like a semitropical plant. It grows best under conditions of high temperature, humidity, and light intensity and with an uninterrupted supply of water and nutrients. Under favorable and stable environmental and nutritional conditions and when pests are under control, the plants grow rapidly and produce heavily. The main stem, laterals, and tendrils grow fast. They need frequent pruning to a single stem and training along vertical wires to maintain an optimal canopy that intercepts maximum light and allows sufficient air movement.

Under optimal conditions, more fruit may initially develop from the axil of each leaf than can later be supported to full size, so fruit may need thinning. Plants allowed to bear too much fruit become exhausted, abort fruit, and fluctuate widely in productivity over time. Excessive plant vigor is indicated by: rapid growth, thick and brittle stems, large leaves, long tendrils, deep green foliage, profusion of fruit, and large, deep yellow flowers. On the other hand, cucumbers are very sensitive to unfavorable conditions, and the slightest stress affects their growth and productivity.

Because fruit develops only in newly produced leaf axils, major pruning may be needed to stimulate growth. The removal of entire weakened laterals is more effective than snipping back their tips.

Important growing parameters

Temperature, general and with special reference to greenhouses

Air temperature is the main environmental component influencing vegetative growth, flower initiation, fruit growth, and fruit quality. Growth rate of the crop depends on the average 24-h temperature the higher the average air temperature the faster the growth. The larger the variation in day night air temperature, the taller the plant and the smaller the leaf size. Although maximum growth occurs at a day and night temperature of about 28°C, maximum fruit production is achieved with a night temperature of 19-20°C and a day temperature of 20-22°C.

During warm weather (i.e., late spring and early fall), reduce air temperature settings, especially during the night, by up to 2°C to encourage vegetative growth when it is retarded by heavy fruit load. This regime saves energy because a 24-h average can be ensured by the prevailing high temperatures and favorable light conditions.

The optimum daily average air temperature is 15-24°C (65-75°F). Optimum temperatures for growth are at night, about 18°C, and during the day, about 28°C accompanied by high light intensity.

To ensure satisfactory stand establishment, **soil temperatures** should be at least 15°C (60°F). The higher the soil temperature, the more rapidly seedlings emerge and the less vulnerable they are to insects and damping-off diseases.

At 15°C (60°F), 9 to 16 days are required for seedlings to emerge. At 21°C (70°F), only 5 to 6 days are required. Even after emergence, cucumbers remain sensitive to cold temperatures. In cold areas, seeds should always be planted late enough to avoid frosts. Exposure to cool conditions will slow growth even if temperatures remain above freezing. Slow-growing seedlings are vulnerable to flea beetles (whose chewing significantly reduces leaf area of young plants). Too high temperatures during flowering decrease pollen viability. Cool and cloudy growing season may cause bitter fruit.

Light, with special reference to greenhouses

Plant growth depends on light. Plant matter is produced by the process of photosynthesis, which takes place only when light is absorbed by the chlorophyll (green pigment) in the green parts of the plant, mostly the leaves. However, do not underestimate the photosynthetic productivity of the cucumber fruit, which, because of its size and color, is a special case. In the process of photosynthesis, the energy of light fixes atmospheric carbon dioxide and water in the plant to produce such carbohydrates as sugars and starch.

Generally, the rate of photosynthesis relates to light intensity, but not proportionally. The importance of light becomes obvious in the winter, when it is in short supply. In the short, dull days of late fall, winter, and early spring, the low daily levels of radiant energy result in low levels of carbohydrate production.

Not only do the poor light conditions limit photosynthetic productivity, but also the limited carbohydrates produced during the day are largely expended by the respiring plant during the long night. The low supply of carbohydrates available in the plant during the winter seriously limits productivity, as evidenced by the profusion of aborted fruit. A fully grown crop benefits from any increase in natural light intensity, provided that the plants have sufficient water, nutrients, and carbon dioxide and that air temperature is not too high.

Relative humidity, with special reference to greenhouses

High relative humidity (RH) generally favors growth. However, reasonable growth can be achieved at medium or even low relative humidity. The crop can adjust to and withstand relative humidity from low to very high but reacts very sensitively to drastic and frequent variation in relative humidity. Its sensitivity to such variation is greatest when the crop is developed under alone can justify installing and operating misting devices.

Carbon dioxide in greenhouses

In relatively high temperatures and light intensity, supplemental carbon dioxide applied at a concentration up to 400 ppm has proved economically useful. Regions with a moderate maritime climate, can more likely benefit from carbon dioxide applied in the summer only. But in regions with a continental climate, the need to ventilate the greenhouse actively throughout the hot summer renders the practice less economical. Apply carbon dioxide during the day or any part of the night when artificial light is supplied. It is economically feasible and highly advisable to use liquid carbon dioxide (carbon dioxide gas liquefied under pressure) because of its guaranteed purity and amenity to accurate concentration control. Liquid carbon dioxide is also preferred because burning natural gas or propane to generate carbon dioxide increases the risk of plant injury from gaseous pollutants, e.g. ethylene.

Seed germination

Seeds germinate and emerge in three days under optimum conditions. During this time seed coat remains tight.

Once cotyledons emerge, roots develop quickly.

Sunlight delivers photosynthates to true leaves and root system.

During the 1st week cotyledons integrity is very important, and if damaged, plants will set back.

Seedlings may recover but they will be weak and susceptible to stresses.

For proper germination, soil temperature must be above 15°C (60°F). If the soil is too cold and wet poor seedling emergence will take place.

Planting

Cucumbers growth season is relatively short, lasting 55-60 days for field-grown varieties, and over 70 days for greenhouse varieties.

Planting dates

Cucumbers are almost always direct seeded. Like most cucurbits, they do not transplant well and transplant costs would be hard to recover. Planting depth is 2.5-4 cm (1-1.5 inches). Too deep delays emergence.

Pickling cucumbers have to be very precise on planting dates so that harvest will coincide with processor needs.

For early crop, container-grown transplants are planted when daily mean soil temperatures have reached 15°C (60°F) but most cucumbers are direct seeded. Early plantings should be protected from winds with hot caps or row covers. Growing on plastic mulch can also enhance earliness.

Spacing

Planting spacing depend on the growth method, variety and harvesting method.

Close spacing increases yields, provides more uniform maturity and reduces weed problems. It also results in shorter fruit with a lighter color.

On the other hand, high plant population requires more seeds and slightly higher fertilizer rates.

Greenhouse plant spacing should provide: 1-2.5/m2 or more per plant, depending on pruning and training system. Recommended density is 33,000 - 60,000 plants/ha.

Figure 1. cucumber greenhouse with trellised plants

Soils

Cucumbers prefer light textured soils that are well drained, high in organic matter and have a pH of 6 - 6.8. Adapted to a wide-range of soils, but will produce early in sandy soils. Cucumbers are fairly tolerant to acid soils (down to pH 5.5).

Greenhouse cucumbers generally grow quite well in a wide range of soil pH (5.5-7.5), but a pH of 6.0-6.5 for mineral soils and a pH of 5.0-5.5 for organic soils are generally accepted as optimum.

When the pH is too low, add ground calcitic limestone, or an equal amount of dolomitic limestone when the magnesium level in the soil is low, to raise it to a desirable level.

Usually the pH in most greenhouse mineral soils is above the optimum pH range (6.0-6.5). A simple, though temporary, solution to a high pH problem is to add peat, without neutralizing its acidity with limestone. Peat also helps to maintain a good soil structure, but it must be added yearly to make up for loss through decomposition.

Soil preparation before field planting

1. Soil fumigation aids in the control of weeds and soil-borne diseases. Fumigation alone may not provide satisfactory weed control under clear plastic.
2. Black plastic mulching before field planting conserves moisture, increases soil temperature, and increases early and total yield. Plastic should be placed immediately over the fumigated soil. The soil must be moist when laying the plastic. Black plastic can be used without a herbicide.
3. Plastic and fumigant should be applied on well prepared planting beds, 2-4 weeks before field planting.
4. Fertilizer must be applied during bed preparation. At least 50% of the nitrogen (N) should be in the nitrate (NO_3) form.
5. Herbicides recommended for use on cucumbers may not provide satisfactory weed control when used under clear plastic mulch on non-fumigated soil.
6. Foil and other reflective mulches can be used to repel aphids that transmit viruses in fall planted cucumbers.
7. Direct seeding through the mulch is recommended for maximum virus protection. Fumigation will be necessary when there is a history of soilborne diseases in the field. Growers should consider drip irrigation with plastic mulch.

Trellising

Cucumber vines can be trained on trellises to save space and improve yield and fruit quality. But the high cost of trellising makes commercial production by this method uneconomical in most cases. Greenhouse cucumbers must be trellised, because the long fruit bend if they rest on the ground.

The major advantages of trellising a cucumber crop include:

- harvesting efficiency
- pest management efficiency
- straighter fruits
- uniform fruit color
- reduction of fruit loss due to soil diseases
- increased yield due to closer rows
- Reduced rate of crooked fruits makes trellising absolutely necessary for Oriental slicing cucumbers

Disadvantages include

- extra cost of trellising materials
- labor to erect the system, dismantle it and training the vines

Figure 2. Different methods of trellising cucumber plants

Greenhouse cucumbers

Flowering habit and fruit set

All European greenhouse varieties produce fruit without pollination. They are gynoecious in flowering habit, and fruits develop without the need for pollination.

Cultural practices

Greenhouse cucumbers grow rapidly under optimum environmental conditions, and fruit production begins 60 - 70 days after seeding. For good production, a temperature range of 24 – 27 Co (75 - 80 Fo) during the day is desirable. While peak daytime temperatures of 29 - 25 Co (85-90 Fo) are tolerable, prolonged periods of high temperatures may adversely affect fruit quality. Night temperatures no lower than 18 Co (65 Fo) will allow a rapid growth rate and earliest fruit production.

Planting and plant growing

Plantings of greenhouse cucumbers are ordinarily started from transplants, but direct-seeding in greenhouse beds may also be practical for late summer or early fall plantings, when the time from seeding to fruiting may not be as critical and prevailing temperatures are warm enough for good seed germination without having to heat the greenhouse.

Transplanting makes more efficient use of greenhouse space, because seed germination and early growth of plants can be confined to a smaller nursery area. Disadvantages to transplanting are the costs of containers and the labor costs of transplanting.

Spacing, training and pruning

The decision on the number of plants to be grown in a given area of greenhouse should be based upon expected light conditions during growth of the crop and also upon the method of pruning the plants. With good sunlight, each plant is allotted about 0.5 m2 (5 feet2) space. Nearly twice that much space may be needed with low light to avoid leaf overlapping and shading by adjacent plants.

Spacing between rows and plants within the row can vary with grower preference. Rows are often spaced 1.2 – 1.5 m (4 - 5 feet) apart, with plants 30 – 45 cm (12 - 18 inches) apart in the row.

Most growers prune their plants by the umbrella system. In this system, all lateral branches are removed as they develop until the plant reaches the overhead support wire. Fruits should not be allowed to develop on the lower 75 cm (30 inches) of the main stem to encourage the plant's rapid vegetative development. Main-stem fruits above that point are allowed to develop at the base of each leaf.

More than one fruit may begin to develop at each node. Some growers thin these to a single vigorous fruit, but it may be more practical to leave all young fruits attached, because it has been observed that more than one may mature.

Fruit thinning

Overbearing can sometimes be a problem. To prevent the plants becoming exhausted and to improve fruit size, control the number of the fruit per plant through selective fruit thinning. This technique is powerful, so use it with great caution. The optimum number of fruits per plant varies with the cultivar and, even more, with the growing conditions. Although, limiting the number of fruits per plant invariably results in premium-priced large fruit, growers risk underestimating the crop's potential or failing to forecast good weather. They may decide to remove too many fruits and thus unnecessarily limit production. Fruit thinning is undoubtedly most useful in the hands of experienced growers who can use it to maximize their financial returns. Fruit to be pruned must be removed as soon as it can be handled, before it grows too large.

Irrigation

Maximum yields and fruit quality will be realized only if the plants receive adequate and timely moisture. Cucumber plants have shallow roots and require ample soil moisture at all stages of growth. When fruit begins setting and maturing, adequate moisture becomes especially critical.

The objective of watering is to maintain a fully adequate supply of water to the plant roots without soaking the soil to the extent that air cannot get to the roots. Do not wait until the plants start to wilt. A good practice is to dig into the soil and judge how much water remains before starting the next irrigation. Regular watering on the same day of the week is unwise. The needs of the plants change daily and seasonally. Water young plants planted in the greenhouse in January or February only once every 5-10 days and then only enough to wet 15-20 cm of the soil. Similar plants growing in June may need 5-10 fold as much water. Let the soil texture and structure determine how much water to add at each application. By examining the soil before watering and several hours thereafter, you can assess the effectiveness of the water application. Depending upon soil type and growth conditions, approximately 25–50 mm of water per week is needed to obtain high quality cucumbers. An irregular water supply, particularly during blossoming and fruit development, can detrimentally affect fruit quality and result in increased nubins or hooked fruit.

The pH of the water is also critical and may need to be adjusted. The target pH of the nutrient solution supplied to the plants should be between 5.5 and 6.0. Nitric, sulfuric, or phosphoric acid are recommended for reducing the pH if it is above 7. If the source water is alkaline due to high bicarbonate concentrations, the pH should be adjusted before the fertilizer salts are added to prevent precipitation. If it is necessary to raise the pH, potassium hydroxide is usually used.

Important: Irrigation timing should not interfere with pollination and should allow surfaces to dry before nightfall.The drip irrigation system is better than most other conventional soil irrigation systems, because it can be used to control crop growth by regulating the supply of water and nutrients. The drip system also allows reduced relative humidity (RH) in the greenhouse, because not all the soil is irrigated and because the system is compatible with the use of white polyethylene film as a light reflecting mulch. In countries suffering from lack of good irrigation water the drip irrigation method has an important feature expressed by high water efficiency leading to very remarkable water savings.

Resources, including irrigation water and energy, are thus used more efficiently with this system. In most cases, use common in-line drippers with a standard flow of 2L/h and one dripper per plant.However, because of the shallowness and extensiveness of the cucumber root system, consider a 4-L/h dripper, which usually results in more lateral movement of the irrigation water. Even better, provide two drippers (2 L/h each) per plant.

Micro-sprinklers, or misters, have also been tried at ground level along the row of plants, with favorable results on root growth, plant vigor, and productivity. However, such irrigation systems, if not properly managed, can easily lead to overwatering and then to disease outbreaks caused by excessive humidity and plant stem wetness. Another alternative is to use lay-flat polyethylene tubes (about 5 cm ID), with small holes spaced 10 cm apart. This system usually applies water to a much larger area than the drip system but does not raise the RH as much as a micro-sprinkler ormicro-mister system. Although fairly inexpensive, the lay-flat tube irrigation system has a short lifespan, which requires its frequent replacement. It is not a good choice for large greenhouses, because the water delivery rates vary significantly along the length of the line (i.e., not pressure compensating). Irrigate the crop up to four times a day, and use the irrigation system to apply fertilizer to the crop.

Fertilization

Greenhouse cucumbers grow quickly and should never be allowed to suffer from lack of water or nutrients. Advanced cucumber cultivation must supply the crop with optimal rates of nutrients throughout the growth cycle in the most efficient manner possible, and without degrading soil and water resources. The nutrient uptake rate by greenhouse cucumbers is very high. One study indicates that cucumbers may require in the range of 28 kg/ha (25 lbs/acre) of nitrogen, 5 kg/ha (5 lbs/acre) of phosphorus, and 40 kg/ha (35 lbs/acre) of potassium per week during peak fruit production. Fertilizer management practices need to assure that plant requirements are satisfied to achieve good yields of high-quality fruit. An important method that can assure that plant requirements are really satisfied is leaf analysis.

Leaf analysis

Visual diagnosis of disorders can be confused by symptoms induced by non-nutritional factors such as disease, pests and chemicals. Leaf analysis can be used to confirm a visual diagnosis. This involves chemical testing of leaves to establish whether specific nutrients are present in plant tissue at normal concentrations. This technique can be used to check the suitability of a fertilizer program and to anticipate the need for nutrient supplements. A disadvantage of leaf analysis is that it is slow, as most laboratories will take at least a week to process samples and report the results back to the grower. In many cases, laboratories do not interpret results or recommend how to remedy the situation.

Nutrigation(fertigation)(the application of fertilizer through the irrigation system) is a popular and efficient **method** of fertilizing field and greenhouse vegetables. Fertilizers are either dissolved in a large holding tank and the solution pumped to the crop, or they are mixed in concentrated stock solutions, which are then incorporated, using fertilizer injectors, into the irrigation water.

Several makes and models of fertilizer injectors are available at varying costs and offer varying degrees of fertigation control. A sophisticated fertilizer injection system capable of meeting the nutrient requirements of a series of crops automatically from the same set of stock solutions was developed may use an IBM-compatible computer to activate a series of dosimetric pumps at varying frequencies for the preprogrammed application of a desired concentration of individual nutrients. It also automatically adjusts the supply of water and nutrients to the crops in accordance with crop and environmental conditions. Introducing drip irrigation and fertigation has made it necessary to consider the fertilizer needs of a crop in terms of the concentration of fertilizer (and therefore the concentration of nutrients) in the irrigation water rather than on the basis of the cropped area.

Furthermore, the recommendations regarding the nutrient content of the fertigation solutions of drip irrigated crops are based mainly on the physiological responses and commercial productivity of the crops. Most earlier recommendations for fertilizer application to traditionally grown crops in soil were based on estimates of nutrient removal by the crop.

Base fertilizers are not normally applied when drip irrigation is used, but peat and lime, may be still needed pre-planting to improve soil structure and adjust soil pH.

Two general examples of fertigation regimes are given, based on leaf analyses done during growth season, but this issue is further elaborated in chapter 3 (dealing with the entire aspect of cucumber mineral nutrition.

Table Nutrigationschedule and rates for cucumber with N:K ratio = 1:2 *

Days after planting			**Daily nitrogen**	**Daily potash**	**Cumulative**	
			nitrogen	**potash**	**nitrogen**	**potash**
			kg/ha		**Lbs/acre**	
(preplant)			28	50	25.0	45.0
0 - 14	0.9	1.8	42	84	37.6	75.2
15 - 63	1.5	3.0	124	220	110.3	196.6
64 - 77	0.7	1.4	135	243	120.1	216.6

Table Nutrigation™ schedule and rates for cucumber with N:K ratio = 1:1 *

Days after planting	Daily nitrogen Daily potash Cumulative					
			nitrogen potash nitrogen potash			
			kg/ha		Lbs/acre	
(preplant)						
8 - 21	1.5	1.5	58	58	52.0	52.0
22 - 63	2.0	2.0	152	152	136.0	136.0
64 - 70	1.5	1.5	168	168	150.0	150.0

* Based on leaf analysis.

Figure 3. Nutrigation™ is a well-established method to increase cucumbers productivity.

Crop rotation

It is recommended not to rotate to crops in the same family during a 3-year cycle for most diseases, but 5 years must be maintained after Fusarium wilt incidence.

Cucumbers Phytophthora blight is the same as buckeye rot on tomatoes. All curucbits get it,although there are differences in susceptibility. Therefore, cucumbers should be rotated with crops other than peppers, eggplants, tomatoes, and other cucurbits.

Specific sensitivities

Cucumbers and other cucurbits have a number of specific sensitivities that may jeopardize yields and fruit quality. Among these sensitivities we can mention general salinity (high E.C. values), sensitivity to specific ions such as calcium, magnesium, sodium, chloride and perchlorate.

Salinity

Salinity damages may be caused by high E.C. ground water, irrigation water, soil or growth medium, and by excessive application of fertilizers.

Figure 4. High salt or E.C.: Leaves appear dull and leathery. A narrow yellow border develops around the leaf margin.

Plants grown under saline conditions are stunted and produce dark green, dull, leathery leaves that are prone to wilting (Figure 2.5). A narrow band of yellow necrotic tissue is often present on leaf edges (Figure 2.6). This can affect leaf expansion, causing a slight downward cupping of the leaf.

After a serious water stress, the oldest leaves may develop a uniform pale green chlorosis and small necrotic areas within the leaf. If water supply is maintained, leaves may only develop a band of pale green tissue around their edges. Plants are likely to wilt in warm weather.

Numerous studies have showed a linear decrease in the yields of cucumbers as the salt concentration of the irrigation water increased. The yield reduction was about 17% for a 1 mmho/cm increase in the E.C.

Figureclearly shows also that cucumber plants are sensitive to saline conditions. Salinity hinders the vegetative growth of the plants. Moreover, salinity severely reduces crop productivity as shown in Figure below.

Cultural treatments to counteract salinity effects are:

- Leach the soil or growing medium with fresh water until the excess fertilizer is removed.
- Adjust the fertilizer program to ensure that rates do not exceed crop needs.
- Use cultivars that are more salinity resistant.
- Apply abundant amounts of potassium fertilizer, other than potassium chloride, as potassium markedly enhances plants ability to cope with salinity stresses.

Chloride (Cl-) toxicity

Normal cucumber growth requires only small quantities of chloride similar to iron, but if the supply is plentiful more is taken up. Chloride toxicity can develop into a serious problem.

Figure 5. Chloride toxicity symptoms on cucumber leaf

Particular attention is required in recirculated hydroponic growth systems. Crop analysis showed that when sodium and chloride were added to the irrigation water, chloride was taken up in much greater quantities by the cucumber roots than sodium. The chloride anion markedly reduces plant vigor and tends to accumulate in the leaves margins, producing a band of pale green tissue around the leaf margin with some leaf-edge scorching and necrosis (tissue death), that stems from concentration rise up to 3% (!!!) (See Figure 2.9.) Such leaves are prone to premature leaf abscission, and reduced photosynthesis activity.

Due to the specific chloride sensitivity of cucumbers, it is highly recommended to avoid fertilizers containing chloride, such as potassium chloride (muriate of potash (KCl), or calcium chloride (CaCl2) in field grown cucumbers. Moreover, chloride-free fertilizers are an absolute prerequisite in hydroponic systems for best cucumber yields.

Perchlorate toxicity

Perchlorate anion (ClO4)-is present in mineral deposits of natural nitrates. Therefore, it is sometimes present in fertilizers produced from those deposits, like potassium nitrate produced in Chile. Perchlorate is a strong acid which decreases the activity of the enzyme RuDP (Ribulose Diphosphate Carboxylase).

When used for the cultivation of greenhouse vegetables it was proved that perchlorate at minute concentrations of 0.3% of the nutrient solution resulted in the "***Bolblad***" syndrome manifested by the following symptoms:

- Leaf curl and malformation developing into partial necrosis.
- Partial opening of the female flowers.
- Marked reduction in yields due to reduced fruit-set.
- Malformed cucumbers.

Glyphosate toxicity

Glyphosate is a very popular wide-range herbicide. When used near cucumber plants, minute amounts absorbed by the cucumber plant are enough to produce damages that include upwardcurling, pale green to yellow younger leaves and severe stunting.

Figure 6. Glyphosate toxicity symptoms on cucumber leaves

Yields

Yields of cucumbers greatly vary according to varieties and growth conditions. The world average is 15 t/ha, while yields of 350 t/ha or more are obtained in modern greenhouses in Europe.

Beit Alpha varieties can yield: 250-450 t/ha and Dutch varieties yield 400 t/ha.

Where yields are 30-60 t/ha under poor cultivation conditions, they can reach 100-300 t/ha under optimal temperature, humidity, light intensity and pollination, achieved under glass).

Harvesting

Cucumber crop matures within 40 - 50 days and harvesting starts 45-55 days after planting.

Harvest during summer to early fall depending on planting time, and variety.

Unless an once-over mechanical harvester is being used, mainly for the pickling industry, fresh consumption cucumbers should be harvested at 2-4 day intervals, when the fruits have reached desired size, to avoid losses from oversized and over-mature fruit.

Important: Over mature cucumbers left on the vine inhibit new fruit set.

Pickling types are harvested when fruits are 5-7.5 cm (2-4 inches) long, and slicing types (for fresh market) - when fruits are dark green, firm, 15-20 cm (6-8 inches) long with a diameter of 4-5 cm ($1\frac{1}{2}$-2 inches).

Fresh-market cucumbersare ready for harvesting when they are about 6 to 10 inches long and 1.5 to 2.5 inches in diameter. The cucumber should be dark to medium green, without any signs of yellowing. On average, 58 to 65 days are required from seeding to maturity, depending on the cultivar and the growing conditions. Fresh-market cucumbers are harvested by hand. Because the individual fruits do not develop and mature consistently, the timing of maturity is not uniform within a field. As a result, fresh-market cucumbers typically are picked between 6 to 8 times over a 3-week period. In some situations, fresh-market cucumbers can be picked up to 12 to 15 times in a season. The number of pickings depends on when the seeds are planted and the supply and demand situation in the market. Price is an important factor in picking. Once prices for cucumbers fall below a certain level, it becomes uneconomical for growers to continue harvesting.

□□□

Chapter 9

Frequently Asked Questions

1. What is Agri-clinics centre?

The agri-clinics are envisaged to provide expert services and advice to farmers on cropping practices, technology dissemination, crop protection from pests and diseases, market trends and prices of various crops in the markets and also clinic services for animal health etc. which would enhance productivity of crops and animals.

2. Who is eligible for getting Agri-clinic training?

Agriculture graduates or graduates in horticulture, animal husbandry, fisheries, and dairy, vet nary, poultry, Pisciculture & other allied activities can apply for a loan under the scheme.

3. What is the duration of the training programme?

Two months training course will be provided free of cost in various State Agricultural Universities and Departments and institutes. Initiated by Small Farmers Agri-business Consortium (SFAC) and coordinated by National Institute of Agricultural Extension Management (MANAGE), Hyderabad,the course comprises entrepreneurship development, business management as well as skill improvement modules in the chosen areas of activity.

4. What are the benefits of Agri-clinic centers?

- Supplement the efforts of Government extension system
- Make available supplementary sources of input supply and services to needy farmers
- Provide gainful employment to agriculture graduates in new emerging areas in agricultural sector

5. What are ventures that can be taken up?

- Soil and water quality cum inputs testing laboratories (with Atomic Absorption Spectrophotometers)
- Pest surveillance, diagnostic and control services
- Maintenance, repairs and custom hiring of agricultural implements and machinery including micro irrigation systems (sprinkler and drip).
- Agri Service Centres including the three activities mentioned above (Group Activity).
- Seed Processing Units.
- Micro-propagation through Plant Tissue Culture Labs and Hardening Units.
- Setting up of Vermiculture units, production of bio-fertilisers, bio-pesticides, bio-control agents.
- Setting up of Apiaries (bee-keeping) and honey & bee products' processing units.
- Provision of Extension Consultancy Services.
- Hatcheries and production of fish finger-lings for aquaculture.
- Provision of livestock health cover, setting up veterinary dispensaries & services including frozen semen banks and liquid nitrogen supply.
- Setting up of Information Technology Kiosks in rural areas for access to various agriculture related portals
- Feed Processing and testing units.
- Value Addition Centers.
- Setting up of Cool Chain from the farm level onwards (Group Activity).
- Retail marketing outlets for processed agri-products.
- Rural marketing dealerships of farm inputs and outputs

6. Who gives the subsidy and to what extent?

The Ministry of Agriculture, Government of India has launched various programmes for setting up of agri-clinics and agri-business centres. These schemes are mainly financed through NABARD and also under Horticulture Mini Mission Project being implemented in some states like Himachal Pradesh, Uttaranchal, Jammu and Kashmir and North-Eastern states. Agricultural graduates are eligible for loan for setting up the agri-clinic and agri-business centers as per provisions of the scheme. The maximum limit for granting loan is Rs. 10 lakh per head in case individual center and Rs. 50 lakh in case of group project. Depending on type of venture, loans can be repaid in 5-10 years as per easy installment plan with the initial moratorium of up to 2 years.

Unemployed youths other than agricultural graduates can take the financial help/ loan from various other schemes being financed/ sponsored by The Ministry of Agriculture, Government of India, National Horticulture Board, State Departments of Agriculture, Horticulture, Animal Husbandry and Fisheries. Horticulture Mini Mission Project with a financial outlay of Rs. 1000 crores being run in some hilly states of the country as mentioned earlier is also providing financial assistance uptoRs. 20 lakhs to start various horticultural ventures. To boost the horticultural activities in the country, the Central Government has given a quantum jump in financial outlay. Thus, agriculture has the real potential to provide the jobs to the unemployed youth in this country. This will increase the agriculture production, provide sustainability to the production and will also earn the country the valuable foreign exchange.

7. What is hand holding support?

Graduates in agriculture and allied subject have been offered a lifetime opportunity for setting up agriclinics and agribusiness centres. In response to newspaper ads by SFAC, 10131 applications have been received by MANAGE. Out of a select list of 20 activities, maximum number of agricultural graduates opted for "Rural Marketing Dealerships of farm inputs and outputs" followed by "Facilitation and Agency of Agricultural Insurance Service" and the third priority was given to "Setting up of Vermiculture Units, Production of Bio-fertilizers, Bio-pesticides, etc."

M.R. Morarka-GDC Rural Research Foundation, having over seven years of experience in Agripreneurship has been recognized as Training cum Resource Centre and to be a member of the "Network of Agri-clinics and Agribusiness Centres by Agriculture Graduates Scheme" of the Government of the India. Over 300 Agripreneurs have already been supported by the Foundation to establish Agribusiness Centres to produce Agri-biotechnology based inputs. At an average investment of Rs. 5.00 lakhs, they have earned an average of Rs. 2.5 lakhs net income in first year itself.

The Foundation has now evolved a comprehensive Agri-entrepreneurial Training design. The hand holding support by the Foundation shall provide an opportunity to conduct detailed feasibility study and approval of business plans. The entrepreneurs would also establish linkages for technical know-how, marketing and term loans during the training itself.

All 25 Agricultural Graduates in First Batch are also being given an offer for joint collaboration with Morarka Foundation to setup an integrated Organic Agriculture-Vermicasting Enterprise.

What is NCPAH?

National Committee on Plasticulture Applications in Horticulture (NCPAH) is constituted in the Ministry of Agriculture (MOA), Govt. of India to focus in a coordinated manner to popularize adoption of various plasticulture applications in horticulture & over all development of plasticulture applications in the country.

What is PFDC?

Precision Farming Development Centres (PFDCs) been established in India to promote "Precision Farming &Plasticulture Applications for hi-tech horticulture" and located in State Agricultural Universities (SAUs); ICAR Institutes such as IARI, New Delhi; CIAE, Bhopal & CISH, Lucknow and IIT, Kharagpur. Presently 22 PFDC have been operating to promote various plasticulture applications in horticulture by undertaking Trials, Demonstrations in state focused crops, Workshops / Seminar, Transfer of Technology through training & awareness programmes, use of Radio / TV programmes, Publications etc in different agriculture climate zones of the country.

Government Schemes?

- NMMI?

  National Mission on Micro Irrigation (NMMI) is a centrally sponsored scheme implemented by the Ministry of Agriculture

(MoA), GoI since 2005-06 to promote use of Drip / Sprinkler Irrigation Systems in the country.

- NHM?

  National Horticulture Mission (NHM) has been launched during the year 2005-06 as a Centrally Sponsored Scheme to promote holistic growth of the horticulture through area based regionally differentiated integrated approach of water management, protected cultivation, nutrition & pest management, post harvest, processing & marketing.

- HMNH

  In order to improve the improve livelihood opportunities and bring prosperity to the North Eastern Region (NER) including Sikkim; Government of India has launched Technology Mission (TM), which is now known as Horticulture Mission for North East and Himalayan States (HMNH). The Mission is based on the "end to end approach" taking into account the entire gamut of Horticulture development, with all the backward and forward linkages, in a holistic manner.

- RKVY?

  National Development Council (NDC), resolved that a special Additional Central Assistance Scheme know as RashtriyaKrishiVikasYojana (RKVY) be launched during 2007. The scheme would focus on agricultural development strategies reoriented to meet the needs of farmers and called upon the Central and State governments to evolve a strategy to rejuvenate agriculture.

- NFSM and ISOPOM?

  The National Development Council (NDC) during 2007 launched a Centrally Sponsored Scheme, 'National Food Security Mission' comprising rice, wheat & pluses to increase the production of rice by 10 million tons, wheat by 8 million tons & pulses by 2 million tons by the end of the eleventh five year plan (2011-12. The National Food Security Mission will have 3 components (i) Rice (ii) Wheat & (iii) Pulses.

In order to provide flexibility to the States in implementation based on regionally differentiated approach, to promote crop diversification and to provide focused approach to the programmes, Centrally Sponsored Integrated Scheme of Oilseeds, Pulses, Oil palm and Maize (ISOPOM) being implemented from 2004

Precision Farming?

Precision farming or precision agriculture is about doing the right thing, in the right place, in the right way, at the right time. The intent of precision farming is to match agricultural inputs and practices as per crop and agro-climatic conditions to improve the accuracy of their applications.

Plasticulture?

Plasticulture represents use of applications of plastics in Agriculture, Horticulture, Water management & related areas. Plasticulture applications offer a multitude of benefits and are considered most important indirect agricultural inputs which results moisture conservation, water saving, reduction in fertilizer consumption, helps in precise application of water & nutrients, controlled environment agriculture is economically viable, plant protection through the use of nets and use of innovative packaging solutions help in increasing shelf-life and during collection, storage & transportation of fruits and vegetables.

Why Plasticulture?

Plasticulture applications are considered the most important indirect agriculture input which results in moisture conservation, water saving, reduction in fertilizer consumption, helps in precise application is economically viable, plant protection through the use of nets and use of innovative packaging solutions help in increasing shelf-life and during collection, storage and transportation of fruits and vegetables.

Different type of Plasticulture Applications?

Broad Classification of plasticulture applications are as follows.

Water management

- Lining of canals, ponds & reservoirs with plastics film

- Drip & Sprinkler Irrigation
- PVC & HDPE pipes used for water conveyance
- Sub-surface Drainage

Nursery Management

- Nursery bags, Pro-trays, Plastic plugs, Coco-pits, Hanging baskets, Trays etc.

Surface cover cultivation

- Soil Solarisation
- Plastics Mulching

Controlled environment agriculture

- Greenhouses
- Shade net houses
- Low tunnels
- Plant Protection nets

Innovative Packaging

- Bins, boxes, leno bags, unit packaging products etc.
- CAP Covers, Controlled Atmospheric Packaging(CAP) & Modified Atmospheric
- Packaging (MAP)

Advantages of plastic in Agriculture?

The unique advantages of plastic over conventional materials are:

- Higher strength/weight ratio
- Superior thermal insulation properties
- Excellent corrosion resistance
- Superior flexibility
- Resistance to most of the chemicals

- Excellent moisture barrier properties
- Favourable gas permeability.
- Smooth surface - resulting in reduction in friction losses
- Excellent light transmissibility
- Helps to enhance shelf-life of the produces
- Better visibility of the produce

Chapter 10

Horticulture Mission

Meghalaya has three factors conducive for the development of horticulture - a large extent of land suitable only for horticultural crops, diversity in agro-climatic factors making a variety of fruits, spices and plantation crops feasible and established tradition of growing horticultural crops making further expansion easy.

Net area sown is only 2.83 lakh hectares forming 12.7 per cent of the geographical area and 7.69 lakh hectares is available in the form of fallows, cultivable waste and miscellaneous tree crops. Most of this land is in the hill slopes and is more suitable for plantation and fruit crops rather than traditional agricultural crops. The State has a great diversity of agro-climatic factors. The average annual rainfall is high at 12000 mm and elevation varies from 150 m to 2000 m mean sea level. As a result, temperature varies from 2° to 36° C. With this diversity, a variety of horticultural crops can be grown.

Given the tremendous potential for the development of horticulture to generate income and employment for the farming community, a much more intensive, focused and holistic strategy, coupled with major investments, is needed. The Mission will take up the task of conceptualizing, designing and executing projects in a systematic and focused manner. Mand for silk is showing an increasing trend; so also the price of silk.

Objectives of Horticulture Mission

The Horticulture Mission has the following objectives

- Expand the area under horticulture by about 36,000 hectares in five years covering nearly 90,000 farmers. This will add 20 % to the existing area under horticulture.
- Adopt holistic approach for the development of horticulture by providing support for irrigation, technology transfer and post harvest management.

- Improve the productivity of existing orchards by rejuvenation of senile plantations, introducing proper management practices and replacing the low grade varieties with high grade.
- Develop farmers' organisations to derive the benefits of higher prices through exporting to outside areas in the country and outside and converting some of the crops into organic.
- Develop cold chain for high value and perishable crops and provide access to the cold chain through development of collection centres at appropriate locations.
- Develop entrepreneurship in processing and marketing of horticulture products.
- Strengthen the existing horticulture farms to produce adequate planting material and convert them as hubs so that they can create spokes and link to the cold chain.
- Encourage the farmers in organic production and assist them in certification.
- Improve the productivity of existing orchards by rejuvenation of senile plantations, introducing proper management practices and replacing the low grade varieties with high grade.

Mission Strategy

The Mission will follow cluster approach to strengthen the existing concentration of crops. For each crop post-harvest management and value chain management will be given emphasis so that additional income and employment will be generated in the State itself.

New technologies in post-harvest infrastructure like grading, packaging, ripening chambers will be introduced for the major crops in areas of their concentrations. As it is difficult to attract huge investment to start large scale processing units, small scale and cottage units will be encouraged and support will be provided for them. Farmers associations will be formed to take care of the management of the post-harvest infrastructure developed.

The Mission will be implemented in a convergence mode for optimising the resource use. MGNREGS, Water Resources Department and Department of Commerce and Industry will be the major players in convergence.

Demonstration of new technologies and crop management practices will be taken up on the fields of progressive farmers who have got the assistance for inputs. The success stories of the farmers will be documented by audio-visual methods and shown to other farmers.

Capacity building will be an important component for all the beneficiaries as well as officials. Farmers' attitudes for the adoption of the practices and their difficulties will be taken into account at the time of introducing new technologies.

The Mission will be co-terminus with the Twelfth Five Year Plan and it will be implemented through mini-missions. The mini-missions are intended to focus on a few activities and discharge them thoroughly. Entrusting all the activities to one agency will not lead to efficient implementation.

Mini Mission I: Area Augmentation

The Mission aims at expansion of 35,731 hectares or 88,293 acres during the period of five years. The major emphasis of the Mission is on the expansion of area under fruits. Area targeted for fruits is 16,270 hectares, accounting for 45.5 per cent of the total area. Next in importance is vegetable group with an area of 11,070 hectares, which forms 31.0 per cent of the targeted area. The targets for plantation crops and spices (including medicinal plants) are 4605 hectares and 3,761 hectares respectively. They account for 12.9 per cent and 10.6 per cent in the additional targeted area respectively.

Cost of Area Augmentation and Expansion (Mini Mission I)

The fruits sub mission has a target of expanding 40,204 acres, covering about 40,000 beneficiaries. The total cost of this sub mission is estimated at Rs. 19,388.64 lakh, out of which Rs.11,631.90 will be provided in the form of assistance and Rs.4850.37 lakh in the form of bank loan. The balance amount of Rs. 2,096.37 lakh has to come as beneficiary contribution, either in the form of labour or in the form of own materials.

The target fixed for vegetables sub mission is 29,107 acres, covering about 30,000 beneficiaries. The total cost of this sub mission is estimated at Rs. 8,957.69 lakh, out of which Rs. 5,520.86 will be provided in the form of assistance and Rs. 3,436.83 lakh in the form of own funds. There will be no bank loan in this case as there is no long or medium term investment involved in it. However, beneficiaries are eligible to take crop loan for the season, but it will not be a part of the Mission.

Plantations, spices, aromatic and medicinal plants will be developed in an area of 26,700 acres and nearly 27,000 farmers will be benefitted under this sub mission. The total cost of this sub mission is estimated at Rs. 9,759.88 lakh, out of which Rs.5,855.93 lakh will be provided in the form of assistance and Rs.1,353.47 lakh in the form of bank loan. The balance amount of Rs. 2,550.51 lakh has to come as beneficiary contribution, either in the form of labour or in the form of own materials. The area augmentation and expansion Mini Mission will cover an extent of 96,011 acres and will benefit about a lakh farmers. The total cost of the Mini Mission is Rs. 38,106.21 lakh, of which assistance is Rs. 23,008.69 lakh and bank loan Rs. 6,203.84 lakh. This amount will be spent in five years with proper sequencing of schemes and coverage of areas.

Mini Mission II: Planting Material

Nursery Development

A nursery for Khasi Mandarin to produce 80,000 budded plants of high quality with indexing will be developed at a cost of Rs. 236.25 lakh. This has to be established in the departmental farm. Two nurseries are proposed- one for Upper Shillong and another for Rongram. At the later part of the Mission, five more nurseries are proposed without indexing facility to be established in the private sector with financial and technical support from the Government.

Planting material for strawberry will be produced in the State by importing the mother plants. The project will be first tested in the government farms and then introduced in the farmers fields. The cost of implementation in the farmers' field will be Rs. 450 lakh, of which Rs. 270 lakh will be assistance and Rs.112.50 lakh bank loan.

It is proposed to establish seven cashew nurseries- five in the private sector and two in the government sector - at a unit cost of Rs. 18.44 lakh. The private nurseries will be assisted with 60 per cent assistance and 25 per cent bank loan.

Support will also be provided to establish five areca nut nurseries and six black pepper nurseries at a unit cost of Rs. 4.45 lakh and Rs. 5.00 lakh respectively.

Planting material for temperate fruits will be available from the existing nurseries and no fresh investment is proposed for this. Production of vegetable is very important for the development of this component. Low cost nurseries will be assisted to produce seed and planting material for vegetables. The unit cost of a nursery is estimated at Rs. 3.00 lakh and 78 nurseries will be supported under this scheme.

The total cost of Planting Material Mini Mission is Rs. 2560.12 lakh, of which 1,914.84 lakh will be provided in the form of Government assistance and Rs. 403.30 lakh in the form of bank loan. A small amount of Rs. 241.98 lakh has to come as own in- vestment by the entrepreneurs. In the case of support to the existing departmental farms, entire amount has to come as assistance and there will be no bank loan or own contribution.

Mini Mission III: Post Harvest Management

Post harvest management is very important for horticultural crops because it will increase the shelf life of the produce. It is estimated that 35 to 40 per cent of the fruits and vegetables valued at Rs. 40,000 crores in India are lost every year due to poor post harvest. Once separated from its sources of water and nourishment, the produce will decay soon in the absence of care. Post harvest hand ling is intended to delay this process as long as possible. Post harvest management, covers pre-cooling, storage, cleaning, grading, packaging and marketing.

The Mission intends to provide integrated cold chain and preservation infrastructure facilities by linking the groups of producers to the processors and market through well- equipped cold chain infrastructure. The infrastructure in the cold chain will consist of sorting, washing, grading and packing infrastructure, multi-chamber cold storages, normal storage, reefer vans, pre-cooling vans and distribution hubs. The Mission identified 13 routes for the cold chain and all these will reach either the border hats or national high way in Assam or Guwahati air port. The cost of the cold chain is estimated at Rs. 781.00 lakh.

Ginger is one important horticulture produce in the State. The produce is sold on the raw form and is subjected to severe price fluctuations. It is suggested that farmers will have storage facility at the point of production for a short period. If the price is very low and storage is also difficult, processing can be taken on a small scale. Hence, the Mission will assist the producer groups to construct 84 ginger go downs at a unit cost of Rs. 4.00 lakh. The scheme can be linked to the central scheme of rural go down construction. Similar is the case with cashew. Since storage is less complex for cashew as compared to ginger, only nine go downs are proposed. Farmers face problems in drying and slicing ginger. Hence, SHGs will be assisted to procure slicer and hydro-extractor for ginger and 550 machines are proposed at a unit cost of Rs. 0.65 lakh.

The State produces a variety of fruits and further expansion will take place in the next five years. Some of these fruits are indigenous and not available in other areas. There are some units in the private sector like Mansan in Ri- Bhoi district. It is suggested that primary processing of the fruits can be taken up at the point of production and transported to the processing units. In order to introduce this new concept, it is proposed to establish six pre-processing units – three in the Khasi Jaintia region and three in the Garo region. However, only one unit in each will be started initially and the other two will be established after examining the performance of these units. It is suggested that the Ranikor area can be selected for initial development and Khadarshnong and Mairang in the later stage. Rongram and Chibinang are identified as suitable locations in Garo Hills. The units may be sanctioned to the successful SHGs or co-operatives and technical guidance and training may be provided by the Department.

The private processing units and the units in the government sector will also be supported to reduce their cost. It is found that transport of empty bottles at a high cost is one of the reasons for high cost of production of fruits juices. To solve this problem, assistance may be provided to the processing units for purchasing pet blowing machine.

Tea Processing

Organic certification is obtained for 100 ha in Ri-Bhoi district, but there is no factory to produce organic tea. A factory with a capacity of 2,500 kg green leaves per day has to be established in the cooperative sector. The cost of this factory is estimated at Rs. 200 lakh, out of which Rs. 150 lakh will be provided as assistance and Rs. 50 lakh as loan.

A large extent of area under tea plantations is neglected in West Khasi Hills as there is no factory in the nearby area. A factory has to be established in co-operated sector at Riangdo at a cost of at Rs.150 lakh, out of which Rs. 112.50 lakh will be provided as assistance and the balance amount of Rs. 37.50 lakh will be the bank loan.

In West Garo Hills, tea is cultivated in an extent of 600 ha and there is a factory under the co-operative sector at Rongram. In the peak season, the capacity of the factory is not sufficient to process the green leaves procured. This is affecting the quality of tea produced. The problem can be solved if two more withering trays are added so that the green leaves can be withered immediately and kept for processing without affecting the quality. The cost of two withering trays of size 40'x 5'is estimated at Rs.5 lakh. This amount can be provided by way of Rs.3.75 lakh assistance and Rs. 1.25 lakh loan.

The following pattern of Ri-Bhoi tea farmers, the farmers in West Garo Hills also want to cultivate tea organically. In fact, they are not using any chemical fertilizers at present. It is pro- posed to obtain organic certification for 200 ha. The total cost of the certification is estimated at Rs. 32.50 lakh.

The cost of cold chain infrastructure is estimated at Rs. 1860.00 lakh and 65% of this amount can be obtained as grant from the MoFPI and the balance amount of Rs. 690.00 lakh has to be invested by the Government of Meghalaya. The total cost of all the components of post harvest management is estimated at Rs. 3715.78 lakh and the expenditure to be incurred by the State government is estimated at Rs. 1739.23 lakh. The remaining amount will come as the grant from the Government Of India and loan from commercial banks.

Implementation Process

Implementation of the Horticulture Mission requires a separate administrative set up because of the massiveness of the Programme. A separate administrative structure will be created at the State, district and block levels. There is also a need to identify the land to be taken up for development and care will be taken to avoid lands that will result in adverse impact on environment.

The following steps will be taken before commencing the plantation activity:

- The implementation will begin with Mass Mobilisation Campaigns to make the people aware of the new areas that can be developed under the Mission. The assistance of NGOs can be availed for the purpose with some lump sum payment for the activity.
- Plantation crops will be developed in clusters so that SHG can be formed for each cluster.
- Beneficiaries in a cluster will form into groups for effective implementation of the Programme.
- It is preferable to have one acre for each beneficiary, but because cluster approach is followed, area more than one acre may sometimes have to be allowed.
- The area identified for a cluster will have water sources. If it is not presently available, it may be developed under convergence with MGNREGS. However, expenditure for bringing water to the individual field can be met from the Mission. The cost of such expenditure will not exceed Rs. 10,000 per acre.
- Based on the estimated cost of cultivation per acre, each beneficiary will be provided 60% of the cost in the form of assistance, 25% in the form of bank loan and the balance is the contribution of the beneficiary. The contribution of the beneficiary will be in the form of labour required for the development of the plantation or own material used.
- After selecting the beneficiaries, they will be provided with the material for fencing and vermin-compost. They have to make the field ready for plantation. The vermin-com- post will be ready and fencing work will be completed. Then planting material will be provided. Plantation activity can be taken up under convergence with MGNREGS so that the beneficiaries need not invest their own funds much.

The Meghalaya State Horticulture Mission will be implemented by the Meghalaya Horticulture Development Agency (MHDA), an autonomous body created for the implementation of the Mission. The Minister of agriculture will be the Chairman of the Agency and the Commissioner and Principal Secretary, Agriculture and Horticulture will be the Deputy Chairman. Principal Secretary, Rural Development will also be a member to facilitate convergence with MGNREGS.

It will have a two tier system of organisational structure, one at the State level and the other at the District level. The Di- rector, Horticulture is the Chief Executive Officer (CEO) at the State level. In view of the heavy work pressure in the Department for implementation of other State and Central Schemes, an additional Director will be appointed as the Deputy Chief Executive Officer (DCEO) to take care of all the technical and administrative matters. However, the DCEO will be of the rank slightly below the director and will have all the powers to take decisions regarding implementation with proper information to the Director Horticulture.

At the district level the DHO will be the nodal officer with separate office and staff for the implementation of the Mission. The District Mission Office will be run by an Assistant Director who will be taken on deputation from the Department with horticulture back ground. In addition to this, there will be two Programme managers and two deputy Programme managers with B.Sc. horticulture. The HDO in the block will be the nodal officer at the block level who will be assisted by 2-4 multiple service providers appointed on contract basis for the period of the Mission. They will possess a minimum qualification of class XII.

Monitoring and Evaluation

The MSHM will organise monitoring in two steps and evaluation again in two steps. The first step in monitoring is con- ducting the base line census survey of all the beneficiaries to obtain their socio - economic characteristics and the lands identified for the development of horticulture. The second step will be the collection of information for the MIS.

Evaluation will be entrusted to a third party and it will be conducted in three stages. The first stage of evaluation will be after completing one year of implementation. This will help to identify the deficiencies in implementation. The second evaluation will be the mid-term appraisal conducted after the completion of second year. It will focus more on the aspects not touched in the first evaluation and also the benefits de- rived by the farmers covered in the first year. The third and final monitoring and evaluation report will be at the end of the fifth year. The samplesize of the evaluation study will be 1000 farmers in the first year, 1500 farmers in the second year and 1500 farmers for the final evaluation.

Management Information System (MIS)

Management Information System (MIS) needs to be introduced in the Horticulture Mission so that all relevant data are collected stored and retrieved instantaneously to facilitate planning, executing, monitoring, and evaluation.

Human Resource Development

Successful execution of various components of the Horticulture Mission will call for systematic building of the competencies of various stakeholders to the required degree. The Mission intends to enhance the capacities of the stakeholders in efficient management of crops, reducing the post-harvest wastage and undertaking primary processing of fruits.

The Horticulture Officers will be trained in the following aspects

- Project management and book keeping
- Participatory rural appraisal methods (PRA)
- Cold chain management

Training of Trainers

Training for the beneficiaries has to be organised at the district level. Active SHGs and Co-operative members may be engaged for training. This will build local capacities and strength- en the organisations. Selected members may be trained along with the Programme managers. They will be trained intensively and material will also be provided. These trainers can be from any district in the region. Thus, two batches of trainers will enlist and provided training so that they can in turn train the beneficiaries.

Creating / Strengthening Training Infrastructure

Since training will be provided at the district level, physical infrastructure will be developed in each district. The gaps in infrastructure will be filled with the help of funds from the Mission. However, expenditure on these items will be kept at a minimum.

Awards to Horticulture Farmers and Incentives to Departmental Officers.

To encourage innovations in management, the Mission will identify progressive farmers and give awards at the end of each year separately to individual farmers, communities and co-operative societies. The criteria for selection will be enhancement of productivity, introduction of new initiatives and management practices.

Scope for Convergence

Two schemes in the MSHM can be implemented efficiently under convergence with MGNREGS. One is planting and fencing of fruit and plantation crops and the other is rejuvenation of cashew orchards. Both these schemes are highly labour intensive and require participation of orchard owners. The machinery needed for rejuvenation will be provided by the MSHM and implementation can be done by the Rural Development Department under MGNREGS.

Majority of the works relating to water catchments, water conservation and water management are handled by the Department of Water Resources. As more than 50 per cent of MGN- REGS works are related to water conservation, the possibility of convergence between MGNREGS, Department of Water Resources and Horticulture Mission will achieve sustainable development of the horticulture sector.

The Department of Commerce and Industries, Government of Meghalaya is implementing National Mission on Food Processing (NMFP). The main objective of the scheme is to increase the level of processing, reduction of wastage, value addition, enhance the income of farmers as well as increase exports. The scheme envisages provision of financial assistance for setting up of new food processing units. It will provide grant in aid of 33% subjective to a maximum of Rs. 75 lakh. It has another scheme of developing cold chain infrastructure. The objective of the scheme is to provide integrated and complete cold chain and preservation infrastructure facilities. Pre cooling facilities at production sites, reefer vans, and mobile cooling units can also be provided under the Integrated Cold Chain Projects. The total cost of implementation, human resource development, M & E and other components is estimated at Rs. 9429.41 lakh. Unlike in the case of the three missions, the entire expenditure has to be borne by the government.

Cost of Post-Harvest Management: Mini Mission III

The total cost of the MSHM is estimated at Rs. 53811.55 lakh for five years. The amount includes the support of Government of India, loan component of commercial banks and private investment, besides the assistance provided by the Government of Meghalaya. The funds for the Government of Meghalaya may be coming from various sources like 13th Finance Commission, RKVY, BRGF or plan funds for gap filling. The assistance of Meghalaya is estimated at Rs. 36,092.17 lakh, which forms two-thirds of the total cost of the Mission. One important component to be included is the rope way at important locations of orange plantations in the valleys to bring the produce out of the valley without causing much waste of the harvested produce. However, detailed study of the existing rope ways and their functioning has to be taken into ac- count before taking up new schemes.

Cost of Implementation, Human Resource Development, M & E and Others

The total cost of the MSHM is estimated at Rs. 53811.55 lakh for five years. The amount includes the support of Government of India, loan component of commercial banks and private investment, besides the assistance provided by the Government of Meghalaya. The funds for the Government of Meghalaya may be coming from various sources like 13th Finance Commission, RKVY, BRGF or plan funds for gap filling. The assistance of Meghalaya is estimated at Rs. 36,092.17 lakh, which forms two-thirds of the total cost of the Mission. One important component to be included is the rope way at important locations of orange plantations in the valleys to bring the produce out of the valley without causing much waste of the harvested produce. However, detailed study of the existing rope ways and their functioning has to be taken into ac- count before taking up new schemes.

❑❑❑

Chapter 11

Khashi Mandarin

Scientific Name		*Citrus reticulata*
Common Name		***Khasi***: Sohniamptra; ***Assamese***:Humoptira/ Komola; ***Bengali***:Komla; ***Manipuri***: Komla
Family		
Importance		
Agroclimatic Conditions	**Climate**	• However, on higher elevations, mandarin orange has been found to loose its sweetness. Mandarin orange and other citrus fruits can be grown well within a range of 2000 mm to 3000 mm of annual rainfall, evenly distributed throughout the year. • Heavy rain at the time of flowering, after a long dry spell, can cause excessive vegetative growth and poor flowering.
	Soil	• Citrus fruits grow in almost any soil. However, the soil best suited for citrus is a well-drained loam or clay loam with a depth of 2.5 to 3.5m and rich in organic matter. • In the north eastern region, mandarin orange has been growing remarkably well in the calcareous rock and acidic soils, producing oranges of excellent quality particularly in Meghalaya • In this region citrus fruit do best on soils with pH range of 5.5 to 7.5.

	Tempera-ture	
Propogation		**By seed** The seedlings, due to polyembryony, come remarkable true to type. Weak seedlings are generally gametic, while those which are uniform in growth and size are believed to arise from vegetative tissues and should be selected for future plantation. **Rough lemon (*Citrus jambheri*):** It is one of the best root stocks for mandarin orange in this region and also in other parts of the country. • This rootstock is most suitable for planting in loose soil. At nursery stage, growth of budded top is slow but after planting in an orchard they grow faster. • Citrus fruits budded on rough lemon produce fruits of an average quality and with comparatively thicker rinds. • Mandarin orange, sweet orange and grapefruit can be successfully propagated on rough lemon root stocks. • This root stock has the ability to withstand Tristeza virus. **Karna khata (*Citrus karna*):** It is another root stock used for propagating various citrus plants in many parts of the country, but is not recommended for north eastern region. **Rangpur lime (*C. limonia*):** • This root stock has proved very successful for raising some sweet orange and mandarin orange varieties in Maharashtra and Karnataka. • This root stock is resistant to Tristeza virus but highly susceptible to exocortis. • It is also recommended for this region till any other rootstock is found to be promising.

Cleopatra mandarin (C. reshni):

This root stock is popular in few citrus producing countries, and has been introduced in our country.

This root stock is resistant to fungal diseases and is tolerant to Tristeza virus.

It is compatible with most of the commercial citrus varieties and makes a smooth bud-union. This root stock may be used in this region after trial.

Trifoliate orange (Poncirustrifoliata): This root-stock can be successfully used for planting citrus fruits in high altitudes. It is capable of imparting hardiness to trees in cold areas and resistant to soil borne diseases.

A root-stock trial at ICAR Complex, Nayabangla, Shillong has been laid out with six different standard root-stocks viz., C. volkameriana, C. taiwanica, Estes rough lemon, Cleopatra mandarin, citrange and rangpur lime using khasi mandarin scion. Preliminary observations showed that rangpur lime followed by C. volkameriana and Estes rough lemon may prove promising.

- In this region a virgin forest soil is generally selected for plantation, hence land preparation starts one year ahead of planting.
- Land preparation includes cutting of forest trees and vegetation, followed by burning and removal of stumps.
- Complete destruction of stumps is essential to avoid termite growth in plantations, especially in hill areas.

Inter-cultivation

- In the north eastern hill region, citrus plantations are seldom put under planned cultivation, and plantations are always kept under sod or raised as mixed crops.
- In the southern slopes of Khasi hills of Meghalaya, the orchards are mostly of mixed types, having other perennial species such as arecanut, jackfruit, banana etc. resulting in a lanky growth of orange trees with a very poor canopy development.

- Undesirable intercrops such as turmeric grown in Jaintia hills in mandarin plantations causes very serious root injury. Mixed plantations and undesirable intercropping are the main reasons for poor yield of existing orchards.

Training and Pruning

- Citrus usually does not require much a proper framework is developed after planting.
- After 6 months to 1 year from the date of planting, 4-5 well shaped branches spread on all the four sides are retained and others cut at their bases 20-25 cm above ground level in case of seedling trees, while 5-7 cm above bud union in case of budded plants.

Nutrient Management

- Similarly, the response of micro-nutrients on the declined trees is also found to be encouraging in this region.
- Therefore, regular spraying of micro-nutrients is essential for normal growth and production of citrus fruits, especially of those nutrients which are deficient in acid soils.
- These micro-nutrients can be applied in April-May along with half dose of nitrogen in spray form at the rate of 2 percent urea, 0.5 percent zinc sulphate, 0.1 percent boric acid, 0.2 percent magnesium sulphate, 0.3 percent copper sulphate and 0.2 percent sodium molybdate.

Fruiting

- Seedling trees of mandarin orange, sweet orange, grapefruit and most other citrus fruits come into bearing within 6 to 8 years of planting. Production in budded plants is allowed to begin after 4-5 years.
- Lime and lemon having a vigorous growth in nature, start bearing after 3-4 years. The bearing span of mandarin orange, sweet orange and grapefruit varies from 25-30 years in budded plants. However, plants from seedlings may produce for a period of 40-50 years.
- The commercial production of lemon and lime covers a span of 12-15 years, and after that they need to be rejuvenated, otherwise quality and quantity gradually decline because of senility.

Citrus Decline

This decline may be due to a single factor or due to the influence of a group of factors, which are described below.

- **Water stress during critical period:** Where irrigation facilities do not exist, severe shortage of soil moisture on the hill slopes during winter and also in pre-monsoon period seems to be and important contributing factor. This problem is more associated with younger plants hence it can be overcome by artificial supply of moisture at least 2-3 times in winter.

- **Sub-soil drainage**: Temporary saturation of soil with free water during prolonged rains seems to cause this problem. However, this may not be a factor in hill slopes.
- **Nutritional deficiencies:** When adequate amounts of plant nutrients are not supplied, there is a gradual decrease in soil fertility which cause decline in citrus orchards. Deficiency of phosphorous, zinc, magnesium, boron, molybdenum, calcium etc. cause decline in production and die-back in this region.
- **Attack of fungal diseases:** In absence of a proper fungicidal programme, especially in dense plantations, trees suffer greatly from the attack of fungal diseases like cracked bark, root-rot and gummosis. They attack both aerial and underground portions of a plant, which disturbs their function and leads to die-back and the resultant severe decline in production.
- **Insect-pests:** Among pests, the shoot and trunk borers are mainly responsible for decline in citrus plantations of north eastern region. They attack shoots, disturb the nutrient movement in plant parts and cause decline in production.
- **Tristeza:** The most important cause of decline is, however, considered to be various viral diseases, occurring throughout the world. Of these, Tristeza has two forms, one causing sudden death of the trees and the other devitalizing them. The latter type is commonly prevalent in this region.
- **Greening:** It is found to be affecting mandarin and sweet orange in this region. The only effective control measure is to use resistant root stocks like rough lemon and Cleopatra mandarin.
- **Plant parasites:** Citrus is a delicate plant which is not able to survive in close competition with wild or natural vegetation like loranthus, lichens, bamboos, wild banana and various wild creepers such as Mikania sp., which

cover the canopy completely. This can be easily controlled by removing these wild growths of plants from individual trees as well as from the entire orchard.

- **Erosion of surface soil:** In the hills, decline of orchards in steep slopes may be associated with erosion of soil by surface runoff due to heavy rains. It causes loss of plant nutrients and exposure of the roots etc. and leads to the gradual decline of fruit plants.
- **Lack of orchard management:** The decline of orange orchards in the north eastern region is apparently due to general neglect and also associated with an absence of scientific cultivation practices. In most of the remote areas, growers are hardly applying any inputs to increase the production of orchards and whatever fruits are available are the gifts of nature. Hence productivity of orange orchards can be enhanced by proper orchard management practices.

The following steps should be taken for rejuvenation of declined orchards.

- The orchards should be cleared and opened by removing other forest species including tall trees, bamboo bushes, wild banana plants etc. Individual trees should also be cleared by removing plant parasites such as loranthus and other wild creepers like Mikania sp. · Pruning of dead and infected branches and application of Bordeaux paste on pruned branch ends.
- By making the base of the trees flat by light harrowing to make them weed free and to facilitate fertilizer application. ·Painting of trunk upto 1 metre from base by Bordeaux paste.
- Application of sufficient amount of manures and fertilizers in the month of March-April along with deficient elements. ·Regular application of insecticides and fungicides as preventive measures. Spraying of Dimecron or roger and 1 percent Bordeaux mixture or 0.5 percent blitox-50.
- Weeds should be cut in the entire orchard area during July-August and mulching should be done with dry leaves near the tree trunks to conserve soil moisture

Diseases and Pests	**Diseases**	**Gummosis** • This is one of the serious diseases of citrus in this region and also in other citrus growing areas. • It is caused by *Phytophthora spp.* namely due to *P. parasitica, P. palmivora, P. citrophthora*etc. this disease commonly occurs at the base of the tree trunk near ground level. • The main symptom is that the bark on the trunk of the affected tree exudes gum slowly and cracks into small chips. **Citrus Canker:** • It is a bacterial disease caused by *Xanthomonascitri.* • Canker attacks leaves, twigs, young branches and fruits. • Spots first become evident on the lower surface, and as the disease advances, the surface of the spot becomes white or greenish and finally ruptures, exposing a light brown, spongy central mass, resulting in a crater-like formation. • Old lesions become corky brown. **Leaf fall and Fruit Rot:** • This is an important disease of mandarin orange in heavy rainfall areas and is caused by *Phytophthorapalmivora.* • The infection starts as water-soaked lesions at the base of the leaves and spreads upwards. • The leaves rot and shed, leaving the tree naked. • The fruits affected by this disease become soft and start rotting. • The basal portion of the fruit stalk becomes weak and fruit drop takes place. • It can be controlled by spraying 1 percent Bordeaux mixture two to three times.

Powdery mildew:

- This disease caused by *Oidiumtingtaninum* is found in mandarin orange. Leaves and tender shoots are mostly affected by the fungus.
- Profuse white powdery growth occurs on both the surfaces of leaves.
- The affected leaves become badly distorted in shape and reduced in size. Dusting with fine powdered sulphur during early morning for a better adhesion on the wet leaves, or spraying with thiovit (0.5 kg in 100 litres of water) can effectively control this disease.

Anthracnose or Wither tip:

- This is caused by *Colletotrichumgloeos-poriodes* on weak plants of lime, citron, lemon and mandarin.
- In affected trees, brown, pimple like cloudy spots occurs on the mature leaves.

Sooty mould:

- In Assam sooty mould is mostly incited by *Capnodiumcitri.*
- Black velvety coating on the leaves, twigs, and fruits is the characteristic symptom of the disease.
- The growth is superficial and can be easily removed from leaf. Under dry conditions the affected leaves curl and shrivel up.

Pink disease (Pelliculariasalmonicolor):

- Generally this disease makes its presence during or after monsoon rains.
- The affected branches are covered with a fine silvery-white film of mycelium. Severely infected bark gets shredded and the wood is exposed.
- Longitudinal cracking and gumming of branches may also occur.

- Leaves turn yellow and finally dry up. Pruning and burning of diseased branches is recommended.
- Mycelium in crotches should be controlled by spraying Bordeaux mixture (5:5:50).

Tristeza:

- It is a virus causes severe damage in some citrus growing areas. Tristeza literally means a sad apance of the tree.
- The tree looks chlorotic and sickly in early stages and dies suddenly.
- In the field, specific symptom of tristeza is honey combing, i.e., a fine pitting of the inner face of the bark below the bud union. Distinct vein clearing and stem pitting can be observed in this disease.

Greening:

- This disease is reported to be caused by mycoplasma like organism (MLO). The characteristic symptom of the disease is yellowing of midribs and lateral veins of leaves.
- In the growing season, upright and weak shoots with bushy growth ap.
- Spraying/drenching with tetracycline (500 ppm) antibiotic gives a temporary control.
- Heating the bud wood in moist air at 47ÚC for 4 hours inactivates the greening pathogen.
- Spraying with BP-101, a new therapeutic at 500 ppm checks the disease.
- The infected trees are required to be removed to avoid further spread.

Pests

There are many insect pests of citrus which cause considerable damage to growth and production viz. stem borer, lemon butterfly, leaf roller, fruit sucking moth, red mites, aphids, thrips, psylla etc.

Citrus trunk borer (*Monohanmus versteegi*):

- It is a serious pest in the north eastern region, most severe in uncared gardens. The larva of the insects bores into the stem, trunk and branches and makes a tunnel inside them.
- It excretes a yellowish-white powdery thread which is quite prominent.

Citrus bark borer (*Indarbela tetraonis and Indarbelaqudrinotata*):

- This is another serious pest of citrus in this region.
- The caterpillar feeds on the bark and thus destroys its translocating tissues.
- It produces a mass of fine slivers of woods and pellets of excreta, mixed with a silky adhesive material on the branches and stems of the trees.
- The control measures are same as that for the citrus trunk borer.

Lemon Butterfly (*Papiliodemoleus*):

- The larvae of lemon butterfly eat the tender leaves making irregular dents on all sides, especially in younger plantations and nursery plants.
- This insect is active throughout the year in this region.
- It lays minute eggs on the margins and tips of fresh tender leaves.

Fruit sucking moth:

- This is a pest occurring in most of the orange growing areas.
- The moth belongs to different genera (Otheris, Achoea, Calpe, Ophideres etc.) which attack the fruits and suck their juice by puncturing them.
- The affected fruits start rotting and fall down from the tree in a few days.

- In order to control this moth, the orchards should be kept free from weeds. The insects are also attracted by tomato fruits.
- They can be controlled by baiting with Malathion (0.05 percent) and producing smoke for two hours after sunset in the orchards.

Leaf roller or leaf miner (*Phyllocnitis citrella):*

- The caterpillars of this insect bore into the tissues inside the leaf and feed upon them.
- Ultimately the leaves get curled outwards and show glistening white galleries or mines on their under surface.
- In case of severe attack, the leaves shrivel and dry up. These damaged leaves get the infection of canker disease very quickly during monsoon.
- This can be controlled by spraying the tree with phosphamidon (0.3 percent), Malathion (0.02 percent) and Metasystox (0.03 percent) at 15 days interval, particularly from June to October.

Citrus aphids (*Toxoptera citricid):*

- These are tiny, green sucking insects found in clusters on tender leaves and branches.
- A black fungus called sooty mould develops on the secretion.
- This can be controlled by a spray of monocrotophos (0.25 percent) and parathion (0.03 percent). This is an insect vector of Tristeza virus.

Citrus Psylla (*Diaphorinacitri*):

- This is a tiny insect gray in colour and attacks the foliage of citrus fruits and sucks the sap and devitalizes the plant.
- The whole tree looks pale with minute specks on leaves in winter season.
- This can be controlled by spraying Malathion (0.05 percent) and parathion (0.025 percent).

- This insect is also responsible for spreading greening diseases.

Mites:

- A number of species such as Tetranychussexmaculatus and T. telarius are found to be prevalent in the north eastern region.
- These mites feed on leaves, makes multiple grey spots, weaken the leaves and cause defoliation.
- For control of red spider mite and oriental red mite, insecticides like monocrotophos (0.05 percent) and Dicofol (0.05 percent) have been found to be effective.

Citrus nematode (*Telenchulus semipenetrans*):

- This nematode has been found in old citrus orchards of Assam.
- It multiplies rapidly in an acid soil with a pH above 4.9 and with a soil temperature range of 25-31°C.
- The symptoms of nematode attack are those of malnutrition and eventually the trees have discoloured, small sized, mottled leaves, died back twigs, small fruits etc.
- Thus, for establishing a new orchard, non-infected nursery stocks should be used.
- Nursery plants can be made nematode free by removing the soil from the roots and submerging them in water at 45°C for about 25 minutes.
- Nursery beds should be regularly fumigated to avoid nematode infections. Chemicals like DBCP (1.2, dibromo-3-chloropropane), nemagon, dasamit etc. can be used for controlling the citrus nematode with reasonable success.

□□□

Chapter 12

Major Criteria's on Which Villages and Crops Selected and Participatory Desicion Support System (PDSS) Formed

Selected Villages for the PDSS

Table. Maximum area coverage in respective selected villages of the Ri-Bhoi District

S.No.	Name of Village	Crop Grown		
			I	II
1	Synghbangla	Cereals	Paddy/rice	Maize
		Fruit	Banana	
		Veg	Pea, beans, potato, Tomato	
		Spices	Ginger	Turmeric
2	Nongkhrah	Cereals	Rice	
		Fruit	Pineapple	
		Veg	Pea, beans, tomato	
		Spices	Ginger	Turmeric
3	Pahambirlum	Cereals	Rice	
		Fruit	Pineapple	
		Veg	Pea, beans, Tomato, potato	
		Spices	Ginger	Turmeric
4	Morkling	Cereals	NA	NA
		Fruit	KM	Lime
		Veg	Potato	Tomato
		Spices	Turmeric	Ginger
5	Pahamrio	Cereals	Rice	
		Fruit	Pineapple	
		Veg	Pea, beans, Tomato, potato	
		Spices	Ginger	

2. Yield Potential

S.No.	Name of Village	Crop Grown		
		I	II	III
1	Synghbangla	Cereals	Paddy/rice	
		Fruit	papaya	
		Veg	Pea, beans, potato,tomato	
		Spices	ginger	Turmeric
2	Nongkhrah	Cereals	Rice	
		Fruit	Pineapple	
		Veg	Pea, beans, tomato	
		Spices	Ginger	Turmeric
3	Pahambirlum	Cereals	rice	
		Fruit	Pineapple	
		Veg	Pea, beans, tomato, potato	
		Spices	ginger	turmeric
4	Morkling	Cereals	NA	NA
		Fruit	KM	Lime
		Veg	Potato	Tomato
		Spices	Turmeric	Ginger
5	Pahamrio	Cereals	rice	
		Fruit	Pinapple, banana, papaya	
		Veg	Pea, beans, tomato, potato	
		Spices	Ginger	

Preference (Growing / Selling / Processing)

	Growing		Selling		Processing				
	Fruit	Veg	Spices	Fruit	Veg	Spices	Fruit	Veg	Spices
V1 Synghbangla	Banana	Pea,beans, potato, tomato	ginger	papaya	Pea, beans, potato, tomato	ginger		tomato	turmeric
V2 Nongkhrah	Pineapple	Pea, beans, potato, tomato	ginger	pineapple	Pea, beans, potato, tomato	ginger	pineapple	potato	turmeric
V3 Pahambirlum	Pineapple	Pea, beans, potato, tomato	ginger	pineapple	Pea, beans, potato, tomato	ginger	pineapple	potato	turmeric
V4 Morkling	Khasi mandarin	Pea, beans, potato, tomato	ginger	Khasi mandarin	Pea, beans, potato, tomato	ginger	Khasi mandarin	tomato	turmeric
V5 Pahamrio	Banana	Pea,beans, potato, tomato	ginger	banana	Pea, beans, potato, tomato	ginger	banana	tomato	turmeric

Germ Plasm Availability

Villages	Fruit	Veg	Spices
V1 Synghbangla	local	Local/ improved	Improved/ hybrid
V2 Nongkhrah	improved/ hybrid	Local/ improved	Improved/ hybrid
V3 Pahambirlum	hybrid	Local/ improved	Improved/ hybrid
V4 Morkling	hybrid	Local/ improved	Improved/ hybrid
V5 Pahamrio	Local(banana)	Local/ improved	Improved/ hybrid

Since for most of the fruits/vegetable and spices available germplasm is local and improved because of this yield of t5hem is less this compled the researcher to introduce the DSS for these crops so that farmers can be acquainted with the improved and hybrid germplasm of the respective crops.

Technology Availability (Out of 7)

Villages	Fruit	Veg	Spices
V1 SYNGHBANGLA	4	3	4
V2 Nongkhrah	3	4	5
V3 Pahambirlum	3	3	3
V4 Morkling	4	3	4
V5 Pahamrio	5	3	4

Since for most of the fruits/vegetable and spices technology available is 3/4 or 5 which is too less then maximum avaible technology i.e. 7 so this compled the researcher to introduce the DSS for these crops so that farmers can be acquainted with the improved technologies for growing fruits vegetables and spices.

Marketing Scope

Village Name	Place				Price				Profit			
	L	M	G	P	L	M	G	P	L	M	G	P
V1 Synghbangla	1	1			1	1				1		
V2 Nongkhra	1	1		1	1	1		1				1
V3 Pahambirlum	1	1		1	1	1		1				1
V4 Morkling	1			1	1			1				1
V5 Pahamrio	1			1	1	1		1				1

L-Loca M-Middleman G-Govt. Agent P-Private Agent

Post- Harvesting – Information

Village Name	Fruit	Veg	Spices
V1 Synghbangla	Banana	Tomato,	Turmeric
V2 Nongkhrah	Pineapple	Potato	Turmeric
V3 Pahambirlum	Pineapple		Turmeric
V4 Morkling	Khasi Mandarin	Potato	Turmeric
V5 Pahamrio	Papaya		Turmeric

Crop Insurance

Village Name	Fruit	Veg	Spices
V1 Synghbangla		NA	
V2 Nongkhrah			
V3 Pahambirlum			
V4 Morkling			
V5 Pahamrio			

CREDIT FACILITY

Village Name	Fruit					Veg					Spices				
	O	P	CO	B	OTH	O	P	CO	B	OTH	O	P	CO	B	OTH
V1 Synghbangla	1					1				1	1				1
V2 NongkhraH	1					1					1				1
V3 Pahambirlum	1					1					1				1
V4 Morkling	1					1				1	1				
V5 Pahamrio	1					1				1	1				1

O-own source P-private money lander CO-Cooperatives B-Banks OTH-Others-ICAR

Criteria for Selection of Village and Fruits Growing in Respective Village

Village-1-Synghbangla

Criteria for selection of village

SYNGHBANGLA: Synbangla has been selected purposively because more than 50 percentages of the farmers here are growing papaya followed by banana. Similarly majority of the farmers are growing tomato followed by beans and capsicum. And most of the farmers are growing ginger followed by turmeric.

Criteria for selection of fruits banana

Banana has been selected because

1. It has maximum area coverage in the village among other fruits followed by papaya
2. It has highest yield potential in the village among other fruits
3. Among other fruits grown in the village banana is preferred in growing and selling
4. Both local and improved (Jahaji) variety of banana are grown by the villagers
5. Out of the seven technologies used as the criteria for selection of the fruit only 4 technologies are used by mos6t of the farmers i.e.

Spacing, planting, fertilizer management and propagation. Technologies like IPM, intercultural operation and processing are not used by the farmers because of the lack of aware ness - this is the reason behind development of the PDSS for banana so that all growing technologies could be transferred to the banana growing farmers

6. For banana there is marketing scope at local market as place but good price is fetched if selling to p-private agent
7. For Banana there is not much post harvesting scope in Meghalaya.
8. Credit for growing banana is provided by own source and private money lander

Criteria for selection of vegetable- tomato

1. Tomato has maximum area coverage in the village among other vegetables followed by Beans
2. It has highest yield potential in the village among other vegetables
3. Among other vegetables grown in the village tomato is preferred for growing and selling
4. Both local and hybrid (Rocky and Baccha) variety of banana are grown by the villagers
5. Out of the seven technologies used as the criteria for selection of the tomato only 5 technologies are used by most of the farmers i.e. Spacing, planting, fertilizer management, propagation and intercultural operation. Technologies like **IPM** and processing are not used by the farmers because of the lack of aware ness - this is the reason behind development of the PDSS for tomato so that all growing technologies could be transferred to the banana growing farmers
6. For tomato there is marketing scope at local market as place but good price is fetched if selling to p-private agent
7. For tomato there is lot of post harvesting scope in Meghalaya.
8. credit for growing banana is provided by own source and private money lander

Criteria for selection of spices-ginger

1. Among farmers of the village more than 50 percentages of the farmers are growing ginger followed by turmeric and spices
2. Ginger has maximum area coverage among the spices.
3. It has highest yield potential also
4. Most of the farmers are growing Nadia Variety.
5. Out of 7 technologies 5 technologies are available to farmers.
6. For ginger there is marketing scope at local market as place but good price is fetched if selling to p-private agent
7. Ginger powder and dried ginger is the post-harvest product of the commodity.
8. Generally own source is used as the credit facility.

Village-2-Nonghkhrah

Criteria for selection of village

NONGKHRAH: has been selected purposively because more than 50 percentages of the farmers here are growing pineapple followed by papaya. Similarly majority of the farmers are growing capsicum followed by beans. And most of the farmers are growing ginger followed by turmeric.

Criteria for selection of fruits pineapple

1. Pineapple has been selected because more than 50 percentages of the farmers are growing pineapple followed by papaya.
2. Pineapple has highest yield potential in the village followed by banana.
3. Pine apple is preferred for growing, selling and processing, generally Kew and queen are grown as the germplasm among them queen is preferred by the farmers
4. Almost 6 technologies are available for growing pineapple.
5. Post harvesting information like making pineapple juice and murrabaha are there for pineapple.

6. Credit for growing pineapple is provided by own source and private money lander.

Criteria for selection of vegetable-capsicum

1. Capsicum has maximum area coverage and highest yield potential in this village.
2. It is preferred by farmers for growing, selling and processing.
3. Out of the 7 technologies available 5 are used by the farmers for growing capsicum.
4. It is sold to local market and middleman but profit could be earned when sold to private agent.
5. Ripe chilli when it turned red, then it is sun dried and powdered and packed.
6. For growing capsicum credit is provided by the private money lander or farmers own source.

Criteria for selection of spices-ginger

1. Among farmers of the village more than 50 percentages of the farmers are growing ginger followed by turmeric and spices
2. Ginger has maximum area coverage among the spices.
3. It has highest yield potential also
4. Most of the farmers are growing Nadia Variety.
5. Out of 7 technologies 5 technologies are available to farmers.
6. For ginger there is marketing scope at local market as place but good price is fetched if selling to p-private agent
7. Ginger powder and dried ginger is the post-harvest product of the commodity.
8. Generally own source is used as the credit facility.
9. Ginger is followed by turmeric (Lagadong)

Village-3-Pahambirlum

Criteria for selection of village

PAHAMBIRLUM: has been selected purposively because more than 50 percentages of the farmers here are growing pineapple followed by banana. Similarly majority of the farmers are growing capsicum followed by beans. And most of the farmers are growing ginger followed by turmeric and chilli.

Criteria for selection of fruits pineapple

1. Pineapple has been selected because more than 50 percentages of the farmers are growing pineapple followed by papaya.
2. Pineapple has highest yield potential in the village followed by banana.
3. Pine apple is preferred for growing, selling and processing, generally Kew and queen are grown as the germplasm among them queen is preferred by the farmers
4. Almost 6 technologies are available for growing pineapple.
5. Local and private market is more suitable for marketing the fruit.
6. Post harvesting information like making pineapple juice and murrabaha are there for pineapple.
7. Credit for growing pineapple is provided by own source and private money lander.

Criteria for selection of spices-ginger

1. Among farmers of the village more than 50 percentages of the farmers are growing ginger followed by turmeric and spices
2. Ginger has maximum area coverage among the spices.
3. It has highest yield potential also
4. Most of the farmers are growing Nadia Variety.
5. Out of 7 technologies 5 technologies are available to farmers.
6. For ginger there is marketing scope at local market as place but good price is fetched if selling to p-private agent

7. Ginger powder and dried ginger is the post-harvest product of the commodity.
8. Generally own source is used as the credit facility.

Village-4-Morkling

Criteria for selection of village

MORKLING: has been selected purposively because more than 50 percentages of the farmers here are growing Khasi Mandarin followed by Assam lemon. Similarly majority of the farmers are growing okra followed by beans. And most of the farmers are growing ginger followed by turmeric and chilli.

Criteria for selection of fruit- KHASHI MANDARIN

1. Khashi Mandarin has been selected because more than 50 percentages of the farmers are growing Khashi Mandarin followed by Assam lemon.
2. Khashi Mandarin has highest yield potential in the village followed by banana.
3. Khashi Mandarin is preferred for growing, selling and processing, generally Kew and queen are grown as the germplasm among them queen is preferred by the farmers
4. Almost 6 technologies are available for growing Khashi Mandarin.
5. Local and private market is more suitable for marketing the fruit.
6. Post harvesting information like making Khashi Mandarin
7. Credit for growing pineapple is provided by own source and private money lander

Criteria for selection of vegetable-Broccoli

1. Broccoli has been selected by the farmers because it has maximum area coverage and highest yield potential among vegetables.
2. It is preferred for growing and selling by the farmers but for processing it is not preferred by the farmers.

3. Pushpa, Aiswarya hybrid varieties are used by the farmers for growing.
4. Out of the 7 technologies available 5 are used by the farmers.
5. Generally local market and private traders are purchasing the broccoli.
6. Broccoli is grown through own source.

Criteria for selection of spices- ginger

1. Among farmers of the village more than 50 percentages of the farmers are growing ginger followed by turmeric and spices.
2. Ginger has maximum area coverage among the spices.
3. It has highest yield potential also
4. Most of the farmers are growing Nadia Variety.
5. Out of 7 technologies 5 technologies are available to farmers.
6. For ginger there is marketing scope at local market as place but good price is fetched if selling to p-private agent
7. Ginger powder and dried ginger is the post-harvest product of the commodity.
8. Generally own source is used as the credit facility.

Village-5-Pahamrio

VILLAGE PAHAMRIO has been selected purposively because more than 50 percentages of the farmers here are growing papaya followed by pineapple. Similarly in vegetables majority of the farmers are growing tomato followed by beans. And most of the farmers are growing ginger followed by turmeric and chilli.

Criteria for selection of fruit- Papaya

1. Papaya has been selected because more than 50 percentages of the farmers are growing papaya.
2. Papaya has highest yield potential in the village followed by banana.

3. Papaya is preferred for growing, selling and processing, generally Kew and queen are grown as the germplasm among them queen is preferred by the farmers
4. Almost 6 technologies are available for growing Papaya.
5. Local and private market is more suitable for marketing the fruit.
6. Credit for growing papaya is provided by own source and private money lander

Criteria for selection of vegetable-Tomato

1. Tomato has maximum area coverage in the village among other vegetables followed by Beans
2. It has highest yield potential in the village among other vegetables
3. Among other vegetables grown in the village tomato is preferred for growing and selling
4. Both local and hybrid (Rocky and Baccha) variety of banana are grown by the villagers
5. Out of the seven technologies used as the criteria for selection of the tomato only 5 technologies are used by most of the farmers i.e. Spacing, planting, fertilizer management, propagation and intercultural operation. Technologies like IPM, and processing are not used by the farmers because of the lack of aware ness - this is the reason behind development of the PDSS for tomato so that all growing technologies could be transferred to the banana growing farmers
6. For tomato there is marketing scope at local market as place but good price is fetched if selling to p-private agent
7. For tomato there is lot of post harvesting scope in Meghalaya.
8. Credit for growing banana is provided by own source and private money lander

Criteria for selection of spices- ginger

1. Among farmers of the village more than 50 percentages of the farmers are growing ginger followed by turmeric and spices
2. Ginger has maximum area coverage among the spices.

3. It has highest yield potential also
4. Most of the farmers are growing Nadia Variety.
5. Out of 7 technologies 5 technologies are available to farmers.
6. For ginger there is marketing scope at local market as place but good price is fetched if selling to p-private agent
7. Ginger powder and dried ginger is the post-harvest product of the commodity.
8. Generally own source is used as the credit facility.

Table. Analysis of the farmers according to the fruit, vegetables and spice growing n=100

Total farmer in each village	No of farmers growing Fruits	No of farmers growing Vegetable	No of farmers growing spices
V1-20(Synhbangla)	20(papaya)	12 (T-B-)	10(G-T-C)
V2-20(Nongkhrah)	20(pineapple)	14(C-B-T)	12(G-T-C)
V3-20(Pahambirlum)	20(pineapple)	16(C-B-T)	8(G-T-C)
V4-20(Morkling)	20(KM)	13(Br-B-T)	5(G-T-C)
V5-20(Pahamrio)	20(Banana)	17(T-B)	7(G-T-C)
Total -Out of 100 farmers	100	72	42

VEG- T = Tomato, B = Beans = Capsicum, Br = Broccoli
Spices- G = Ginger = T = Turmeric, C = Chilli

After analysis of the table it is found that the those farmers who are growing fruits all of them are not not growing vegetables while it is possible for all of them to grow vegetables as intercropping because they can take at least one vegetable during monsoon season/summer season–here also we fins the technology/knowledge gap among farmers through multistory cropping. Farmers could grow vegetable under the fruit trees it will double their income and out of the 100 fruit growing farmers only 42 are growing spices while it is possible to grow Ginger, turmeric and chilli as intercropping with the fruits – here come the PDSS to provide them know -how for multistory cropping.

Table. Crops and Respective varieties grown by the farmers of the selected villages of Meghalaya.

S.No.	Crop	Variety
Fruits		
1	Pineapple	Kew, Queen
2	Banana	Jahaji
3	Khashi Mandarin	Khashi Mandarin/local variety
4	Papaya	Local and hybrid-Washington, madhu-bindu, solo,
5.	Assam lemon	local
Vegetables		
1	Tomato	Local, Hybrid-Rocky, Bocchia
2	Broccoli	
3	Capsicum	California wonder
4	Okra	Pusha sawani
5	Cucumber	Malini
Spices		
1	Ginger	Nadia
2	Turmeric	Lagadong
3	Chilli	Local

Here we found that for some crops e.g. papaya, banana and Assam lemon under fruit crops farmers are growing local germplasm because of the lack of awareness about hybrid germplasm so through PDSS we could provide them information and seed of hybrid germplasm available for the fruit crop. In case of vegetables same is there with cucumber and okra here also through PDSS we can empower the farmer with the hybrid germplasm.

Standard cultivation practices related to fruits, vegetables and spices are provided as package of practices to the farmers.

Technology available with farmers with relation to fruits, vegetables, and spices

	Fruits					**Vegetables**					**Spices**	
Technologies	**P**	**K.M**	**A.L.**	**B.**	**P.**	**T**	**BR**	**C**	**O**	**C**	**G**	**T**	**C**
1. Propagation	1	1	1	1	1	1	1	1	1	1	1	1	1
2. Spacing	1	1	1	1	1	1	1	1	1	1	1	1	1
3. Planting	1	1	1	1	1	1	1	1	1	1	1	1	1
4. Integrated Nutrient Management (INM)	0	0	0	0	0	0	0	0	0	0	0	0	0
5. IPM	0	0	0	0	0	0	0	0	0	0	0	0	0
6. Intercultural Operation	1	1	1	1	0	0	0	0	0	1	1	1	1
7. Processing	0	0	0	0	0	0	0	0	0	0	0	0	0

1 = Tech available
0 = Tech not available
P = PINEAPPLE
T = Tomato
BR = Broccoli
CU = Cucumber
O = Okra
Ca = Capsicum
KM = Khashi Mandarin
A.L. = Assam Lemon
GI = Ginger
T = Turmeric
C = Chilli
B = Banana
P = Papaya

□□□

Chapter 13

Mandarin

Mandarin orange is the most important commercial fruit crops of NE region. It is well known for its quality fruits in the market of North eastern region.

Cultivar: *Khasi* Mandarin

Soil and Climate

Well-drained medium deep soil is to be selected for orchard. Soil having pH of 5.5 to 6.5 is considered most favourable.

Mandarin thrives well in sub-tropical to semi- temperate climates up to 1000msl. It requires high humidity, warm summer and

Propagation

Seed: Seeds should be sown in the nursery immediately after extraction (within 1 week) into a depth of 1.5-2 cm at 1Ox5 cm distance. Seedlings are replanted in secondary bed or polythene bags of size 5" x8" (containing equal part of soil, sand and FYM) at 4-6 leaf stages in polyhouse for speedy and uniform growth. Seedlings for planting should be strong, healthy, and uniform in growth having a height of 60-90 cm and free from pest and diseases. Rootstocks are also raised as above mentioned technique.

Budding

Time: February-March or in July-August

Method: 'T' - budding

Rootstock: Rough lemon and Rangpur lime

Soft wood grafting

Time: July-August

Method: Wedge grafting

Scion selection: Previous season 3-4 months old shoot of 2-4 mm thickness with 3-4 healthy buds of 8 to 10 cm long round shape should be selected.

Rootstock: Rough lemon and Rangpur lime (90-120 days old)

Pit preparation

The size of the pit is 0.75 x 0.75 x 0.75 m and pit refilled with upper 30 cm soil along with 15-20 kg FYM, 100g urea, 100g MOP, 300g SSP and 50g Chlorpyriphos dust or granule. The pits are filled about 15 cm above the ground level.

Spacing

The spacing between the two plants should be 5 x 5 m apart.

Planting time

Spacing nting is June to August. If there is no rain after planting, light irrigation should be given. Bud/graft union should be kept at least 15 cm above the ground level.

Manure and Fertilizer

Depending upon the age of tree, the following manure and fertilizer schedule is followed.

Chemical	Application	I	11	III	IV	V	VI
FYM,kg	Feb-March		5	10	15	20	25
Urea, g	March	100	150	200	250	300	35
SSP,g	March	100	200	300	400	500	60
MOP,g	March	50	100	150	200	250	30
Lime, kg	Feb.		1	2	3	4	5

Type of manure/ Time of Age of tree (Year)

Repeat the same dose of urea, SSP and MOP during June-July and again in September- October. Therefore, requirements of grown up trees may be N-480g, P2Os-290g and K2O- 540g along with 25kg FYM per tree per year. The fertilizers are applied around the basin under the canopy of the tree, 20-25 cm away from the main stem. Besides, 2-3 spray of Zinc Sulphate (0.4%) + Magnesium Sulphate (0.2%) + Copper Sulphate (0.3%) or Multiplex (mixture of micro-nutrient) @2.5m11litre of water should be given during flushing period.

Irrigation

Irrigations are to be given only during dry periods, the first being at planting time, subsequent irrigations can be given at 15-20 days interval during December to March.

Intercultural operation

Weeding should be done frequently at monthly interval, mulching with paddy straw/farm grass or black polythene can also use to control weeds.

Intercropping

Additional income can be obtained by growing suitable intercrops such as French bean, rice bean, cow pea, black gram and other vegetable crops from mandarin orchards during pre-bearing stage (1-5 years).

Control of fruit drop

Control of fruit drop can be achieved by spraying with 2,4-D or GA3 @15 mg/L + Urea (10 g/L) +Benomyl(1glL) in April, May and September or Spraying of Planofix @ 1 mll5 L of water in the month of March-April and August-September to minimize the fruit drop.

Insect/Pests

Leaf miner: Caterpillars feed on newly emerged leaf tissues forming zigzag shrinking streak like galleries.

Citrus scale: All parts except roots, infested by scales. Due to attack the leaves turn yellow, bristle like and fall off. Fruits mottled and shrivelled shoots and twigs withered.

Aphid: Nymphs and adults suck the sap from newly emerged leaves, tender parts and flowers.

Mealy bug: The nymphs and adults suck the sap from plants.

Monocrotophry. It can be controlled easily by hand picking and killing of larvae as well spraying of Chlorpyriphos@ 2 ml/1.

Citrus Trunk borer: Grubs bore and feed on the bark, making tunnels inside the trunk. To kill the trunk borer grubs, clean the bored hole of the infested plant with iron wire and insert a cotton swab soaked in Dichlorvos/ petrol or inject 5 ml of Dichlorvos (2.5 ml llitre) or petrol and plug with mud. Collection and destruction of trunk borer adults during May-June by shaking the branches 2-3 times at 10 days interval may also help in controlling the pest population.

Diseases

Damping off: Cotyledons of newly emerged seedling rotted near the ground.

Spraying and soil drenching with Bavistin@ 2.5g/litre of water is effectively control the disease.

Powdery mildew:A whitish powdery growth is visible on young leaves and green parts. Spraying of Sulfex@ 2.5 gllitre of water during flush period is recommended to control the disease.

Scab: The corky lesion on fruits, leaves and young branches identifies the disease.To control the scab, spraying of Bavistin@ 2g/ litre of water in April, June and September is recommended.

Twig blight: Plant exhibits drying of twigs and small branches from growing tip.

The affected portion should be cut and pasted with Bordeaux paste. Spraying with 1 %Bordeaux mixture or Copper oxychloride@ 2.5 g/L found effective.

***Phytophthora* rot:** The first indication of the disease is exudation of gum from the bark of the stem. The bark cracks open and in the later stage dries up. Drenching of root with 1 % potassium permanganate solution followed by 1 % Bordeaux mixture may save the plant from declining.

Harvesting and yield

Fruits should be harvested when they attain full size, develop attractive colour with a TSS of more than 9^0Brix. Fruits are ready for harvesting during the month of November- December. From 6-year-old tree about 40-50 fruits may be harvested.

❑❑❑

Chapter 14

Okra / Ladys Finger

Cultivation of Okra (*Abelmoschus esculentus*)

Plant Characte-ristics

- It is a tropical direct sown vegetable with duration of 90-100 days. This crop belongs to the family Malvaceae.
- A well-drained deep soil is ideal for its growth. Optimum temperature range is 25-30 ^oC.

Health Benefits of Okra

- The superior fiber found in okra helps to stabilize the blood sugar by curbing the rate at which sugar is absorbed from the intestinal tract. Okra's mucilage binds cholesterol and bile acid carrying toxins dumped into it by the filtering liver.
- Okra helps lubricate the large intestines due to its bulk laxative qualities. The okra fiber absorbs water and ensures bulk in stools. This helps prevent and improve constipation. Unlike harsh wheat bran, which can irritate or injure the intestinal tract, okra's mucilage soothes, and okra facilitates elimination more comfortably by its slippery characteristic. Okra binds excess cholesterol and toxins (in bile acids). These, if not evacuated, will cause numerous health problems. Okra also assures easy passage out of waste from the body. Okra is completely non-toxic, non-habit forming, has no adverse side effects, is full of nutrients, and is economically within reach of most individuals unlike over-the-counter drugs.
- Okra fiber is excellent for feeding the good bacteria (probiotics). This contributes to the health of the intestinal tract.
- Okra is a supreme vegetable for those feeling weak, exhausted, and suffering from depression.

- Okra is used for healing ulcers and to keep joints limber. It helps to neutralize acids, being very alkaline, and provides a temporary protective coating for the digestive tract.
- Okra treats lung inflammation, sore throat, and irritable bowel syndrome.
- Okra has been used successfully in experimental blood plasma replacements.
- Okra is good for summer heat treatment.
- Okra is good for constipation.
- Okra is good in normalizing the blood sugar and cholesterol level.
- Okra is good for asthma. Okra's vitamin C is an antioxidant and anti-inflammatory, which curtail the development of asthma symptoms.
- Okra is good for atherosclerosis.
- Okra is believed to protect some forms of cancer expansion, especially colorectal cancer.
- Eating okra helps to support the structure of capillaries.
- Some information shows that eating okra lowers the risk of cataracts.
- Okra is good for preventing diabetes.
- Okra protects you from pimples and maintains smooth and beautiful skin. We understand the reason why Cleopatra and Yang Guifei loved to eat okra.

Varieties

- **Green/light green fruited:** Pusa Sawani, Pusa Makhmali, IARI Selection 2, Kiran, Salkeerthi.
- **Red fruited:** Co-1, Aruna.
- **Yellow vein mosaic resistant/tolerant:** Arka Anamika, Arka Abhay, Susthira (all green fruited).

Seeds and Cultivation

- The three main planting seasons for Okra are February-March, June-July and October-November.

Seed Rate

- The seed rate is 8.5 kg/ha for the summer crop sown in February-March and 7 kg/ha for kharif crop.

Storage of Seeds

- Packing of okra seeds in polythene cover (700 gauge) increases the storage life up to 7 months.

Planting

- For kharif crop, sow the seeds at a spacing of 60 cm between rows and 45 cm between plants. For the summer crop, soak the seeds in water for 24 hours before sowing and give a spacing of 60 x 30 cm.

Water Management

- Give pre-sowing irrigation, if soil is not moist enough. During summer, irrigate at intervals of 2 to 3 days.

Nutrient Management

- Apply FYM or compost as basal dose @ 12 t/ha. At the time of sowing, apply N, P_2O_5, and K_2O @ 25, 8 and 25 kg/ha. Another 25 kg N per ha may be applied one month after sowing.

Weed Management

- Conduct weeding regularly and earth up rows during rainy season.

Pest Management

- The important pests are jassids, fruit and shoot borer and root knot nematode.
- Against jassids, use quinalphos, fenthion or fenitrothion as foliar sprays each at 0.05%.
- For controlling fruit and shoot borers, remove all drooping shoots and damaged fruits. Spray carbaryl 0.15% at intervals of 15 to 20 days. For controlling aphids, apply dimethoate 0.05%.
- For the control of nematodes, apply sawdust or paddy husk at 500 g/plant or neem leaves or Eupatorium leaves at 250 g/plant in basins one week prior to planting and water daily.
- The effect of this treatment persists up to 75 days after sowing in summer season· Application of Bacillus macerans or B. circulans (1.2 x 106 cells per pit) before sowing is recommended for the control of root knot nematode (ad hoc recommendation).
- For the control of pests, application of carbofuran granules at the rate of 0.5 kg ai/ha or phorate at the rate of 1 kg ai/ha at seeding followed by need-based application of foliar insecticides has been recommended.
- The application of granules is recommended only at the time of seeding.
- In general, insecticides of plant origin may be used, as far as possible.

Diseases **Yellow vein mosaic**

Management

- This is a common disease in okra, which shows vein clearing and vein chlorosis of leaves.
- The yellow network of veins is very conspicuous and veins and veinlets are thickened. Fruits become small and yellowish green in colour.
- White fly (Bemisia tabaci) and leaf hopper (Amrasca biguttula biguttula) are vectors of this virus.
- Hence, their control is very important.
- Use of resistant varieties like Arka Anamika, Arka Abhay and Susthira, and destruction of host weeds (Croton sparsiflora and Ageratum sp.) are also effective.

Harvesting

- The crop comes to harvest 45 days after sowing.
- Since the pods get matured very quickly, the selection of pods with maximum size but still tender is the point to be considered.
- A total number of 15-18 harvests can be made

Average Yield 10-15 tonnes/ha.

❑❑❑

Chapter 15

Papaya

Scientific Name *Carica papaya*

Common Name Papita (hindi), sohkynphor (khasi)—(Garo)

Family Caricaceae

Importance

- Papaya is a common man's fruit, also known as melon tree or papaita.
- it is excellent source of vitamin A and has good amount of vitamin C, Minerals (Fe, Ca and P) and carbohydrates.
- Unripe green papaya is used as vegetable.
- Papain prepared from the dried latex of immature fruits is used in many ways in pharmaceutical industries, tannin industry, cosmetics; tenderization of meat etc.
- Papaya is a popular fruit famous for its high nutritive and medicinal values.
- It comes early in bearing than any other fruit crop, produces fruits in less than a year and the production of fruits is quite high per unit area.
- Papaya is cultivated more or less on a commercial scale in the foothills and plain valleys of all states of the north eastern region. As per statistics available 3,670 hectares yield 47,280 tonnes of papaya annually.
- It is the fourth important crop of this region. Among the hill states, Mizoram has

		the largest area under this crop, followed by Tripura and Manipur, While in production Manipur contributes the maximum, followed by Tripura and Mizoram.
		• Papaya is a native crop of Mexico, and was introduced in India in the 16th century. Now it has become popular all over India and is the fifth most commercially important fruit of the country.
Agro climatic condition	**Climate**	• Thrives well in tropical climate.
	Soil	• Prefers a rich, well-drained soil.
	Temper-ature	• susceptible to waterlogging around the trunk • Low temperature and frost limits the crop growth. • Optimum temperature requirement: 22-26°C.
Varieties	**Improved**	**CO_1 and CO_2** • These two improved varieties have been released by the Tamil Nadu Agricultural University. • CO_1 is a selection from Ranchi variety and it is ideal for table purpose. • CO_2 is suitable both for table purposes and papain production. Both varieties are famous for their dwarf stature.
	Hybrid	**Washington** • It has purple coloured leaf-stalks, purple rings on the stem and yellow flowers. • It also comes fairly true to type from seed. • Fruit is round to ovate in shape, and with excellent sweet taste and flavour. • It is grown in Maharashtra, Gujarat and also in this region. **Honey Dew or MadhuBindu** • It has good flavour and taste of fruit with less number of seeds.

- The bearing is heavy with long fruits in clusters.
- Honey dew trees are medium sized.

Coorg Honey Dew

- Selection from honey dew variety made at IIHR.
- This variety produces no male plants.
- Its plant bears fruits heavily, low on the trunk.
- Its fruit size is medium to larger (2 to 3.5 kg), shape long to oval, central cavity big, flesh orange coloured and keeping quality moderate.
- This variety is grown in the north eastern region.

Solo

- This is a famous variety from Hawaii islands.

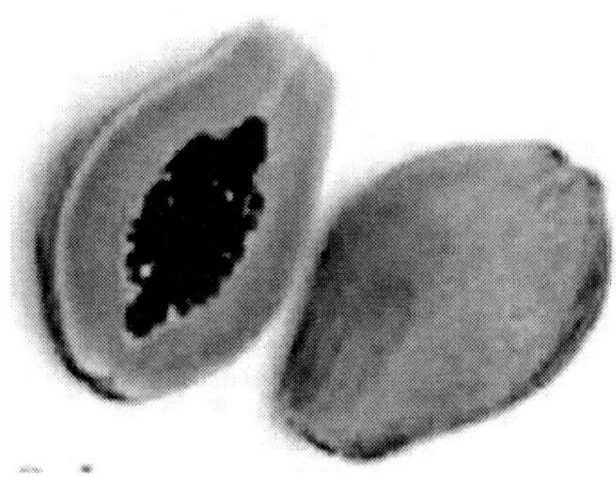

- The fruits are small (0.5 to 1 kg) and pear shaped.
- This variety produces no male plants and requires a uniformly warm and moist climate.

Ranchi

- It performs well under north Indian conditions and is highly valuable in respect of yield and quality.

- It bears large fruits on plants of a medium height.

Barwani

- It is a popular variety in central India, large fruits
- It is susceptible to leaf curl virus disease.

Haflong

- It is a popular variety on north eastern region.
- It produces fruits of excellent quality and good size on tall plants. There are several other new varieties such as Pusa Delicious, PusMajestic.

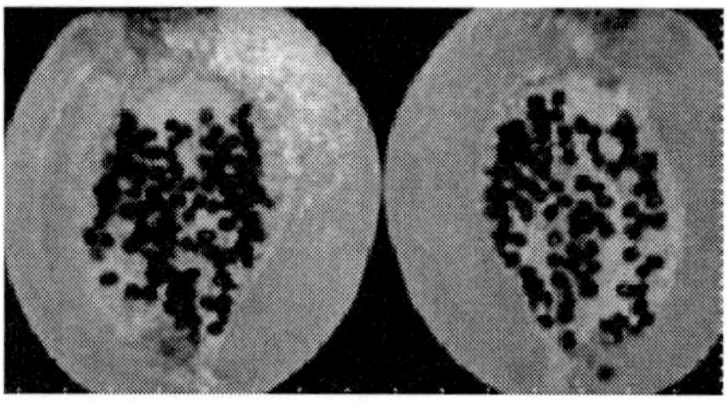

Method of propagation

- Only by **seeds.**
- Seed rate - 250 g/ha
- Vegetative propagation by mound layering is also possible.
- Tissue culture

Nursery rising

- Seeds are sown in raised seedbeds of 2 m x 1 m made 15 cm above the ground level or in polythene bags. A mixture of sand, leaf mould and dried FYM is spread over the seedbed.
- Seeds are sown 2-3 cm deep at a distance of 5 cm (P-P) X 15 cm (R-R).
- Seed beds should be watered daily,
- Seedlings raised in polythene bags can withstand water stress better than that raised in seedbeds.
- Polythene bags of 20 cm x 15 cm size and 150 gauge thickness are used.

- They are filled with a **mixture of FYM,** soil and sand in equal proportions.
- Two seeds are sown in each bag and keep only one seedling after germination.

Planting

Time of planting

- Best time for raising papaya seedlings - February to March.
- Two months old seedlings are transplanted in the main field in May-June at a spacing of 2 m x 2 m.

Planting method

- Pits of size 50 cm x 50 cm x 50 cm are taken and filled with topsoil. Male plants are removed as soon as they flower and the female and hermaphrodite plants are retained.
- In pure female plantations, one male plant is retained for every 10 female plants.
- Seedlings are shaded to protect them from excessive sunlight until they establish.
- In hermaphrodite or monoecious types male plants may not be required.

Nutrient Management

- Organic manure (FYM) may be applied @10 to 25 kg / plant / year at the onset of southwest monsoon in basins around the plant
- Each papaya plant should also be supplied with 40 g N, 40 g P_2O_5 and 80 g K_2O at bimonthly interval.
- Split doses

Intercultivation and Intercropping

- Two hand diggings, one in the beginning of the rainy season and another in January-February are necessary.
- When papaya is grown as the main crop, vegetables can be profitably cultivated as

intercrop for about six months from planting of papaya seedlings.

Water management

- The crop should be irrigated in summer.
- The ring system of irrigation is better for papaya than the basin system because the ring system prevents irrigation water coming into contact with the stem, thus preventing collar rot.
- Irrigate the young papaya seedlings about once or twice a week in irrigated orchards whereas for older trees bearing fruits, irrigate every 15 days or so.
- drip irrigation system is the best one for papaya

Plant protection/IPM

Diseases

Damping-Off

(*Rhizoctonia solani*):

- This is a disease of young seedlings.
- Lesions are seen on the stem at or just above soil level.
- The stem becomes watery and shrinks, followed by death of the plant.

Control:

1. **Cultural measures:** Well-drained soil should be used for planting and the crop

should not be excessively irrigated.

2. **Organic measures** Before sowing the seeds should be treated with fungal culture of *Trichoderma viride*(3-4 g/kg of seed)
3. **Chemical measures**
   - This can be prevented by sterilizing the soil of the seedbed with 2.5 per cent formaldehyde solution and covering it for 48 hours with newspapers or polythene sheets.
   - This treatment is given 15 days before sowing.
4. **Biological measures**
   - Incorporation of smoke-water into water agar medium inhibited growth of Pythium with a complete inhibits on of mycelium growth
   - Observed for smoke-water concentrations of 3.5% and higher.

Collar rot or stem rot: Waterlogging and poor drainage are the chief contributing factors.

Control:

1. **Cultural measures**

   Improving drainage in the orchards. Infected fruit on the trees and those that have fallen to the ground should be removed.

2. **Organic** can be controlled with the virgin soil technique. Virgin soil (soil in which papaya has never been grown in before) is placed in planting holes about 30 cm in diameter and 10 cm deep with a mound about 4 cm high. Roots of papaya plants are protected by the virgin soil during the susceptible stage, and become resistant to the pathogen when they extend to the infested soil.
3. **Chemical measures**

   Application of Bordeaux paste on the stem and soil drenching with Bordeaux mixture are control measures
4. **Biological measures**

   Pseudomonas fluorescens-against Pythium aphanidermatum responsible for papaya root rot

Anthracnose :

(*Colletotrichum gloeosporioides*):

The disease prominently appears on green immature fruits. The disease symptoms are in the form of brown to

black depressed spots on the fruits. The initial symptoms are water-soaked, sunken spots on the fruit. The centers of these spots later turn black and then pink when the fungus produces spores.

Control:

1. **Cultural measures**

   The affected fruits should be removed and destroyed. The fruits should be harvested as soon as they mature.

2. **Organic:**

   Castor oil for the control of anthracnose was shown when 5% and 10% (v/v) were used in the assays resulting in reduced mycelial growth. Fungal sporulation was strongly inhibited at 10% (v/v) concentration of essential oil.

3. **Chemical measures**

   - To control, spray Bordeaux mixture 1 per cent.
   - Spaying with Copper Oxychloride (3 g/litre of water) or Carbendazim (1 g/litre of water) or Thiophanate Methyl (1 g/litre of water) at 15 days interval effectively controls the disease.
   - Fruits for exports should be subjected to hot water treatment or a fungicidal wax treatment.

4. **Biological measures**

   Trichoderma species against *Colletotrichum gloeosporioides* on papaya fruits.The best strains *in vitro* were *Trichoderma longibrachiatum* and *Trichoderma harzianum.*

Insects **Red spider mite (*Tetranychusurticae*):**

Spider mites usually extract the cell contents from the leaves using their long, needle- like mouthparts.

Control :

1. **Cultural measures**:

   burn the weeds and papaya remains after harvesting, summer weather deep ploughing.

2. **Organic measures**

3. **Chemical measures** Affected leaves may be clipped off and destroyed. Dusting with Sulphur or spraying with lime Sulphur or dimethoate 30 ec@1.5ml/l water may be used for control.

4. **Biological measures Predators:**

   Anthocorid bugs (Orius spp.), mirid bugs, syrphid/hover flies, green lacewings (Mallada basalis and Chrysoperlasp.), predatory mites (Amblyseiusalstoniae, A. womersleyi, A. fallacies and Phytoseiulus persimilis), predatory coccinellids (Stethoruspunctillum), staphylinid beetle (Oligota spp.), predatory cecidomyiid fly (Anthrocnodaxoccidentalis), predatory gall midge (Feltiellaminuta), Predatory thrips etc.

Aphid (*aphis gossypii and myzus persicae)*

- Infesting tender shoots and under surface of the leaves.
- Curling and crinkling of leaves
- Stunted growth
- Development of black sooty mould due to the excretion of honeydew

1. **Cultural measures**
   - Soil solarization
   - Timely sowing should be done.
   - Field sanitation, rogueing.
   - Destroy the alternate host plants
   - Apply manures and fertilizers as per soil test recommendations.

- Plant tall border crops like maize, sorghum or millet to reduce white fly and aphid's population.
- Crop rotation with graminaceous crops
- Adopt ecological engineering by growing the attractant, repellent, and trap crops around the field bunds.

2. **Chemicalmeasures**

   Nymphs and adults suck sap from the leaves and transmit mosaic virus disease.

Control

Weeds should be removed which are additional host. Spraying of 250 ml malathion 50 EC in 250 liters of water controls the aphid.

3. **Biological measures**

   Parasitoids:*Aphidiuscolemani, Aphelinus* sp.,

   Predators: Fire ant, Robber flies, Big-eyed bug (Geocorissp), Earwig, Ground beetle, Cecidomyiid fly, Dragon fly, Praying mantis, Lacewing, Ladybird beetle, Spider etc

Root knot nematode

(*Meloidogyne* sp.):

Damage symptoms

- Formation of galls on host root system is the primary symptom

- Roots branch profusely starting from the gall tissue causing a 'beard root' symptom
- Infected roots become knobby and knotty
- Plants wilt during the hot part of day, especially under dry conditions and are often stunted

Control

1. **Cultural measures:**remove all the undecomposed roots, debris etc. of the soil. Follow crop rotation as a preventive measure.
2. **Organic measures** : Application of the neem cake@2 tones/ha tends to reduce the incidence

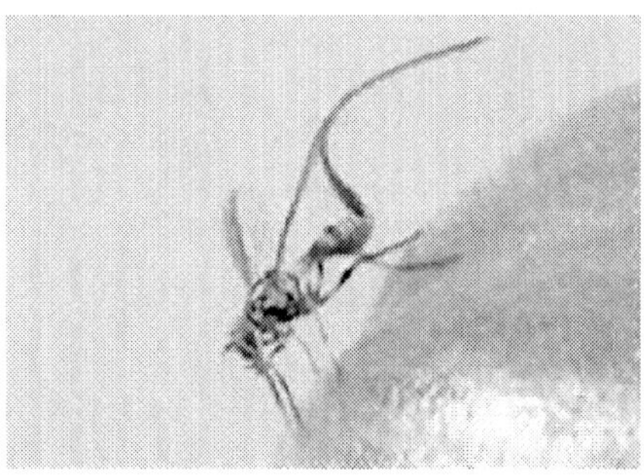

3. **Chemical measures** Addition of the carbafuran3G@4kg/ha controls the nematode
4. **Biological measures**

Verticitlium chlamydosporium

Fruit fly

Damage symptoms

- The female punctures outer wall of mature fruits with the help of its pointed ovipositor and insert eggs in small clusters inside mesocarp of mature fruits.
- On hatching, the maggots feed on fruit pulp and the infested fruits start rotting due to further secondary infection

1. **Cultural measures** :
    - Bagging can be an effective control measure for the fruit fly
    - Sanitation is also important in the control of the papaya fruit fly.

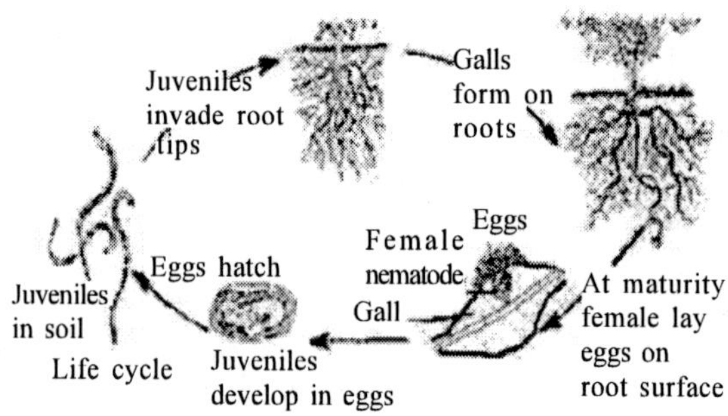

2. **Organic measures:**

   Application of the neem cake@2 tones/ha tends to reduce the incidence

3. **Chemical measures:**

   Addition of the carbafuran 3G@4kg/ha controls the nematode

4. **Biological measures**

   - **Parasitoids:** Opiusfletcheri, Fopiusarisanus, Diachasmimorphakraussi
   - **Predator:** Ants

Marketing

- Both raw and ripe papaya are being marketed in local market and shillong or other nearby markets Processed for papain extraction For local markets optimum temperature of 20°C was found both for ripening and satisfactory storage for two weeks.
- Storage below 10°C has been known to cause chilling injuries greater in mature- green than ripe papayas.
- Chilling injury symptoms include pitting, blotchy coloration, uneven ripening, skin scald, and increased susceptibility to decay.
- More than half ripe papayas (at lest 50% yellow) can be stored at 4-10°C without developing symptoms of chilling injuries.
- Exposure of papayas to temperatures above 30°C for more than 10 days result in heat injury. The symptoms include uneven ripening, blotchy ripening, poor color, abnormal softening, surface pitting and accelerated decay.
- For exports of papaya, hot water treatment is given by dipping the fruits for 30 minutes at 42°C immediately followed by a 49°C dip for 20 minutes.
- **Quick cooling to 13°C** after heat treatments minimizes heat injury.
- **Controlled-atmosphere (CA) storage (2% O and 5-10% CO) at 10°C** has been found beneficial for 2 2 delayed ripening, firmness retention and avoid chilling injuries. Transport for local markets the fruits are transported in trucks while for distant markets it should be preferably sent through railways.

- Transportation through railways is faster and economical.

Processing and Value addition

Harvesting

- A good crop may fail if harvesting of fruits is not done properly due to perishable nature of fruits.
- The fruits should be left on the tree until they mature fully.
- Usually the fruits are harvested when they are full size, light green with a tinge of yellow at apical end.
- On ripening, fruits of certain varieties turn yellow but some of them remain green. When the latex ceases to be milky and become watery, the fruits are considered suitable for harvesting.
- While picking the fruits from the trees, care must be taken that these are not scratched and are free from any blemishes, otherwise there will be attack of fungus and the fruits may start decaying during marketing.
- Packing: Since papaya fruits are easily perishable, care should be taken to wrap the individual fruit in paper and finally packed in wooden crates surrounded by soft material all-round and saw dust especially on the bottom.
- Storing: Reduction in temperature or treatment with retardants reduces the duration for ripening and enhances shelf life.
- Post-harvest treatment of fruits with silver nitrate or cobalt chloride extends the shelf life without affecting the palatability.
- Papaya fruits, at colour turning, can be stored at 7°C, which will have normal ripening.
- Shelf life of fruit is also extended by storing at 13°C with 1.0 to 1.5 percent oxygen or at 10 % CO.
- Waxing of fruit and storage under low pressure (LP) has also been successful in reducing the disease incidence and increasing the shelf life of papaya.

Yield

- The number of fruits harvested per tree per year varies from 25 to 30. Fruits showing streaks of yellow colour are harvested.

- Although papaya trees bear flowers and fruits continuously for many years, it is not economical to retain the trees after 2.5 to 3 years

Year Approximate yield	(kg/ha)
1st	150-200
2nd	200-250
3rd	75-100

Extraction of papain

- Papain is an active enzyme present in the latex or milky secretion of papaya plants and immature fruits.
- Half to three-fourth matured fruits (about 70 to 100 days from fruit set) are preferred for papain extraction.
- Tapping of fruits can be done early in the morning by giving longitudinal skin depth incisions (3 mm) on the surface of the fruits from the stalk end to tip.
- Stainless steel blades or knives or bamboo splinters are used for incising papaya fruits.
- The milky latex is collected in arecanut spathes or aluminium or glass vessels.
- The incisions are repeated in two or three subsequent occasions at 3 to 4 days intervals.
- The latex collected in this way is dried in the sun or in an artificial drier at 50-55°C.
- A small quantity of potassium metabisulphite is added to the liquid latex to extend the storage life of papain.
- The dried latex can be stored in airtight polythene or glass containers for a period of six months. Tapped fruits are equally tasty as untapped fruits, although impaired in appearance.

❑❑❑

Chapter 16

PDSS-What, Why and How

A Theoretical Framework for Development of Participatory Decision Support System for Horticulture Crops

Decision support system research and its applications evolved significantly over the 1970 and 1980s, and DSS was considered as one of the most popular areas in information systems during the time period. Diverse DSS were developed to support decision makers at all levels in an organisation including systems that could support problem structuring, operations, financial management and strategic decision making, even extending to support for optimisation and simulation. Ever since they were created in the 70s, Decision Support Systems (DSS) have been a great source of help with different management problems such as the optimization of travel times in airlines or train companies, medical diagnosis, business management, natural resource management, agriculture and forestry. The application of DSSs to farm management involves a range of opportunities and challenges. The latest years has seen mankind confronted with the problem of food security worldwide. Issues such as declining profitability of agriculture, climate variability and increasing concerns over the environmental impacts of farming pose complex challenges for farm management. DSS is based on the search for technology that can make agricultural systems science more accessible and useful for guiding management of production systems. Agriculture farmers and scientists faces a great challenge in effectively managing information in order to improve the economic and operational efficiency of operations, reduce environmental impact and comply with various documentation requirements. Interactive computer programmes that utilize analytic methods for developing models to help decision makers formulate alternatives, analyse their impacts, and interpret and select appropriate options for implementation. Participatory DSS development involves local actors and researchers cooperating as active co-experimenters engaged in joint learning. Participatory DSS development integrates researchers, developers and users perspectives to clarify objectives and foster co-learning. In order to meet this challenge, the flow of information between decision processes must be analysed and modelled as an output and outcome.

Application of ICT in Agriculture

Economic development Application in planning and market information for agriculture Procedures, provides mobile advisory to farmers. ICT services also provides price for casting of different crops which helps the farmers in deciding which crops to grow to fetch the remunerative prices. Community Development Applications, Research: ICT helps in quick data collection and rapid feedback to the problem faced by the farmers and ready access to the on line literature e.g. research papers of various on line journals and reference materials of various e-books likewise in education ICT provides on line teaching and learning, access to waste and diverse study material through internet. Providing education to remote areas through T.V.channels. Providing Small and medium enterprise development, New media networks, Providing decision support to public administrators for improving planning and monitoring of development programmes, Improve services to citizens, and bring in transparency, Empower citizens through access to information and knowledge .Help in training to improve the functioning of developmental organization and expand Employment opportunities, Market Linkage etc.

Overview of PDSS

A. **Meaning and Concept of PDSS** Participatory Decision Support System (PDSS) is an interactive, flexible, and adaptable computer based information system that utilizes decision rules, models, and model base coupled with a comprehensive database and the decision makers own insights, leading to specific, implementable decisions in solving problems that would not be amenable to management science models. Thus, a PDSS supports complex decision making and increases its effectiveness. -The Skeleton of Information System Building an interactive computer programs that utilize analytical methods, such as decision analysis, optimization algorithms, programme scheduling routines, and so on, for developing models to help decision makers formulate alternatives, analyse their impacts, and interpret and select appropriate options for implementation.

B. **Approach to development of PDSS** There is no universally accepted taxonomy of DSS either. Different authors propose different classifications. Using the relationship with the user as the criterion, Haettenschweiler. Differentiates passive, active, and cooperative DSS.

1. A passive DSS:

   A passive DSS is a system that aids the process of decision making, but that cannot bring out explicit decision suggestions or solutions.

2. An active DSS:

   An active DSS can bring out such decision suggestions or solutions.

3. A cooperative DSS:

   A cooperative DSS allows the decision maker (or its advisor) to modify, complete, or refine the decision suggestions provided by the system, before sending them back to the system for validation. The system again improves, completes, and refines the suggestions of the decision maker and sends them back to her for validation. The whole process then starts again, until a consolidated solution is generated.

4. **One more approach according to author is there i.e.participatory approach**: in this approach participation of the farmer or respondents is ensured while designing the participatory decision support system. Opinion of the farmers about problems faced by them with relation to growing of horticulture crops is collected with the help of schedule .Then decision support system is designed problem oriented.e.g. Some disease is faced by pineapple growers by the farmers of Meghalaya data about the disease are taken by the PDSS developer then appropriate and specific solution is recommended while developing the PDSS.

C. **Type of DSS**the important types of decision support systems are listed as follows

1. A model-driven DSS:

   A model-driven DSS emphasizes access to and manipulation of a statistical, financial, optimization, or simulation model.

Model-driven DSS use data and parameters provided by users to assist decision makers in analysing a situation; they are not necessarily data intensive.

2. A communication-driven DSS:

   A communication-driven DSS supports more than one person working on a shared task; examples include integrated tools like Microsoft's NetMeeting or Groove.

3. A data-driven DSS:

   A data-driven DSS or data-oriented DSS emphasisaccess to and manipulation of a time series of internal company data and, sometimes, external data.

4. A document-driven DSS:

   A document-driven DSS manages, retrieves and manipulates unstructured information in a variety of electronic formats.

5. A knowledge-driven DSS:

   A knowledge-driven DSS provides specialized problem solving expertise stored as facts, rules, procedures, or in similar structures. Our aim is to develop a data-driven DSS for agriculture. In this paper we studies the factor that are taking a part to develop the DSS.

Framework of PDSS

It has been suggested that one of the main contributions to the poor uptake of previous agricultural DSS has been the failure of many developers to pay attention to users, their needs and requirements, and to the environment in which they operate. The proposed framework divided into six main subsection. These sub-systems are in cycle form. The system can be more improved in the process of decision making if the system is validated by requirement analysis and specification. If it is not then it should be reconstructed until the appropriate model is generated. -The Proposed Framework of PDSS Building.

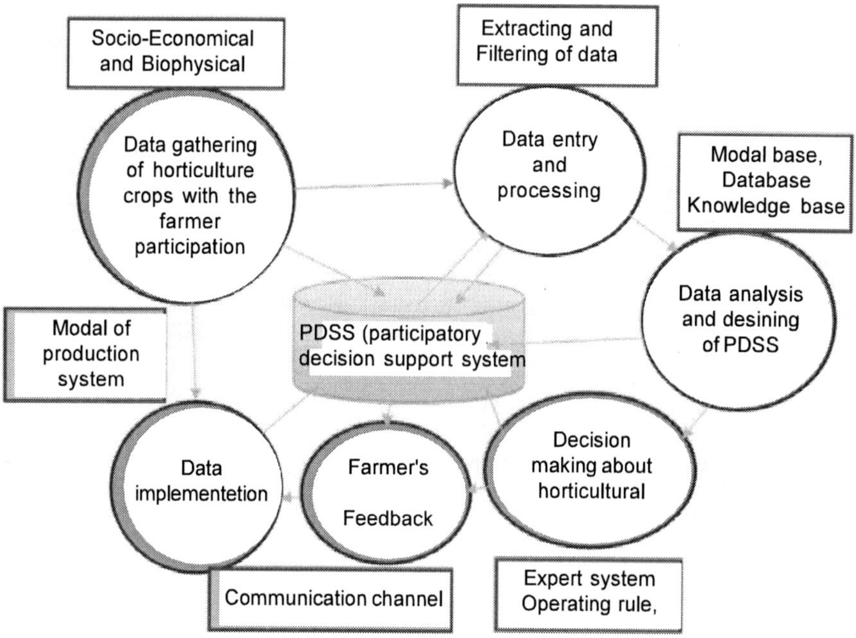

Model of participatory decision support system

A. Data Gathering

Many of Agriculture DSS failed due to unproven and unstructured knowledge and information about the horticulture. It has been found that one of the main contributions to the poor uptake of previous agricultural DSS has been the failure of many developers to pay attention to users, their needs and requirements, and to the environment in which they operate .Many system development methods use the notion of task to describe the user's interaction with the system. Tasks and their components are identified as part of the requirements analysis process and the system's ability to support the users tasks, physical or mental are a measure of a systems success. This environment is characterized by low and variable funding, poor communication between end user and developer, and variable levels of product development skill .Many researcher found different kind of tools to find out the concerning information gathering. Mainly three types of tools were found as it described below.

1. Participatory Research: PR involves: forms of inquiry where researchers and the researched population form collaborative

relations in order to identify and address mutually conceived issues or problems through cycles of action and research .PR can be define by collaborative or collegiate relationships between researchers and stakeholders. This approach is being used for the development of the participatory decision support system for horticulture crops of Meghalaya, cooperation of khasi speaking person is taken while collecting data to know exactly what is the problem faced by farmers. Horticulture crops are taken over agronomical crops because these are more remunerative second hilly terrain of the Meghalaya which support horticulture crops better then the agronomical crops. PR is better over other methods because in this method data about problems faced by farmer for growing horticulture crops are gathered and solution is being provided to them.

2. Questionnaire/Schedule Approach: The models these systems contained were capable of producing information that the user required but the way the system had been designed prevented the user from accessing that information.

3. Workshop: work- shops for idea generation a useful way of tapping the imagination and intuition of crop protection specialists when trying to come up with new options for pest management. They also suggest that their system and decision analysis techniques are best employed in the context of interdisciplinary workshops.

B. Data Processing

The agriculture have been surround by many factor which involves technical (GIS, communication, computer) and non-technical fact (climate, soil, weather). The horticultural knowledge base in particular is growing rapidly, as science improves its understanding of biological and chemical processes, and the quantity of data can be overwhelming. Filtering and assimilating this mass of information is a formidable task particularly given the limitations of human processing. Unlike a computer, we cannot manipulate many things in our heads at one time and we are known to make errors in logical thinking. To store the data related to the processes of interest in the Participatory Decision support system, both spatial and feature related as well as time series data. This section of proposed model involves the registration of measurement into database and their sub-sequent processing, retrieval and storage. The economical and biophysical data (soil, water, forecasting) are kept in the form of information and this information will keep in the form of database after filtering. The information must be according to the boundary of problem and information gathering.

C. Analysis and Design

In the analysis process each function should be analysed separately. There are certain analytical items related relative model will be designed for each analytical item. The design of the model is the key of the whole system. For participatory decision support, relative professional knowledge will also be attached with the models. The models used to infer the state of the system so that reasonable decision alternatives can be formulated. Analysis models and tools should be designed to predict response and provide forecasts. Using data from the Data Collection System, and historical and GIS data needed to calibrate GIS models for the better response of result. The input to this stage is a set of sub-tasks, ranked questions and information sources. The task now is to translate these into a system specification. A system specification is defined as a document that clarifies a client's requirements in an unambiguous form by re-specifying the requirements and distinguishing between system functions and the constraints the system developer has to work under.

D. Decision Making Process

Making the decision could be defined as integration of result produced at stages of process with computer, human logic and integration of previous developed model. " Gathered and merges the conclusions from knowledge-based and numerical techniques and the interaction of users with the computer system through an interactive and graphical user interface". An expert system is defined as a computer program designed to model the problem solving ability of a human expert. Now a days many researcher shift their intention to develop the expert system based PDSS. It is also defined as a system that uses human knowledge captured in a computer to solve problems that ordinarily require human expertise. Consisting of a set of rules and user- supplied data which interact through an inference engine, an expert or knowledge-based system is able to derive or deduce new facts or data from existing facts and conditions. Expert-system shells and programming languages have become widely available allowing users to define databases and rule sets. Most of the horticultural data have geographic attributes, GIS is an important tool for agricultural analysis. It is very important to include GIS into the DSS for regional agricultural management; however, it does not mean that the system should be developed on professional GIS, Multi-objective modelling methods have been used for several decades to determine the trade-offs between various objectives in these problems. Due to the conceptual difficulties involved in using multi objective models (i.e., selecting criteria, specifying satisfying values, and evaluating trade-offs), several researchers have developed multi objective decision support tools which meet two of the three requirements of a DSS. The methods differ in the type of information they request, the methodology used, the sensitivity tools they offer, and the mathematical properties they verify. Indeed, practical applications of the multi criteria approach are hindered by the ambiguity of choosing one particular method among all those available. Each method may potentially lead to different rankings, and the choice of a methodology is subjective and dependent on the decision makers pre-disposition. The operating rule can be defined as outline and rule of DSS in which the system will have to operate. The communication channel is defined as the medium of massage and technology passing between the scientist and farmer to educate them about the system operation and operating rules. ICT, e-mail, mobile device, EDI and many other medium can be used to massage passing. Training of farmers mainly end user could lead to better performance of DSS System.

E. Farmers feedback

Farmer feedback is essential to make the DSS a participatory for this feedback is taken in a schedule / questionnaire. Then Solution of heir problem is incorporated while developing the DSS. This will make the the DSS participatory and problem oriented.

F. Decision Implementation

Decision implementation System have made for disseminating decisions regarding horticulture under normal conditions, and flood warnings, weather forecasts, and disaster response in affected and specified areas. This block represents the actual decision made by the decision maker (single or group) based on the results of the multi criteria analysis, Expert system and GIS.

Some basic data as field location, field soil type or some constant data can directly retrieved from the system database while others that are more transient such as soil, water content, biophysical content have to be input by user. The system then will use those data, supported by analysis models and domain knowledge, to give an understandable decision result. In simple words, it can be defined as formulation of actions to be implemented in solving a specific problem". The above development begins with the collection and processing of data, followed by analysis of various horticulture problem domains and then analysis with Expert and interpretation along with decision maker. The outcome model of decision maker will lead to formulation of plan. In practice, the above stated system development process is not in the step-by-step process but in cyclical process which involve all the sub-part of system. All the sub-part process i.e., entering data in process (information gathering), analysis being performed, and decision being taken in almost in the continuous fashion.

Conclusion

With the increasing of population throughout the world the need for diversify the horticulture production, there is a definite need for improvement of management of world horticulture resource. This could be solved by more and more application of ICT and their involvement in every stage of production from soil identification to till production. In present scenario, every part of the world is conducting the survey to gather empirical information on vegetables, fruits and condimentsand other agriculture resource. Good decision making is always a critical. Good planning always lead to good decision making. Planning and development process of horticulture PDSS constitute a multi-complex problem which is very difficult to be solved, if it is not faced thoroughly. The interdependence between various activities on one hand and service other hand are making the planning very difficult. The decision making processes demand from producer (information)/ decision maker a qualitative evaluation and other problem that they are facing. Our proposed model involves social-economical, GIS, Expert system help to decision maker to develop. Our process model could lead to make horticultural participatory DSS more appropriate and lead to expected horticulture solution model.

❑❑❑

Chapter 17

Strawberry

Common Name

- Strawberry

Scientific Name

Fragaria sp.

Introduction

- Strawberry (Fragariasp.) is a native of temperate regions, but varieties are available which can be cultivated in subtropical climate.
- In India it is generally cultivated in the hills. The strawberry is the most widely adapted of the small fruits. Strawberries are grown throughout Europe, in every state of the United States, as well as in Canada and South America.
- The wide variation in climates within these regions and the wide adaptation of the strawberry plant permit harvesting and marketing, the fruit during greater part of the year.
- Strawberry is a delicious fruit taken fresh in several ways.
- It also makes excellent ice cream and Jam on account of its rich aroma, and is also a good source of vitamin C.
- It is a soft and a highly perishable fruit.

Agroclimatic Climate condition

- Strawberry thrives best in temperate climate.
- It is a short day plant, which requires exposure to about 10 days of less than 8 hours sunshine for initiation of flowering.
- In winter, the plants do not make any growth and remain dormant.

- From the standpoint of response to length of the light period, strawberries are placed in two groups: (1) varieties that develop flower buds during both long and short light periods, the overbearing varieties and (2) varieties that develop flower buds during the short light periods only, most commercial varieties.
- Strawberry requires a well-drained medium loam soil, rich in organic matter.

Soil

- The soil should be slightly acidic with pH from 5.7 to 6.5. At higher pH root formation is poor.
- The presence of excessive calcium in the soil causes yellowing of the leaves.
- In light soils and in those rich in organic matter, runner formation is better.
- Strawberry should not be cultivated in the same land for a number of years. It is preferable to plant it in green manured field.
- Alkaline soils and soils infected with nematodes should be avoided.

Temperature

- The exposure to low temperature during this period helps in breaking dormancy of the plant.
- In spring when the days become longer and the temperature rises.
- The plants resume growth and begin flowering.
- The varieties grown in milder subtropical climate do not require chilling and continue to make some growth during winter.

Varieties

- A large number of varieties are available.
- For the hilly areas, varieties Royal Sovereign, Srinagar and Dilpasand are suitable.

- Some of the introductions from California, such as Torrey, Toiga and Solana may prove even more successful.
- The variety found successful in Bangalore has been named Bangalore and which has performed well at Mahabaleshwar also.
- For the north Indian plains, Pusa Early Dwarf which has dwarf plants, large firm wedge-shaped fruits, has been recommended.
- Another variety with rich aroma but softer fruits is Katrain Sweet.

Propa-Gation

- Propagation is done by means of runners that are formed after the blooming season.
- The plants may be allowed to set as many runners as possible but not allowed to set any fruits.

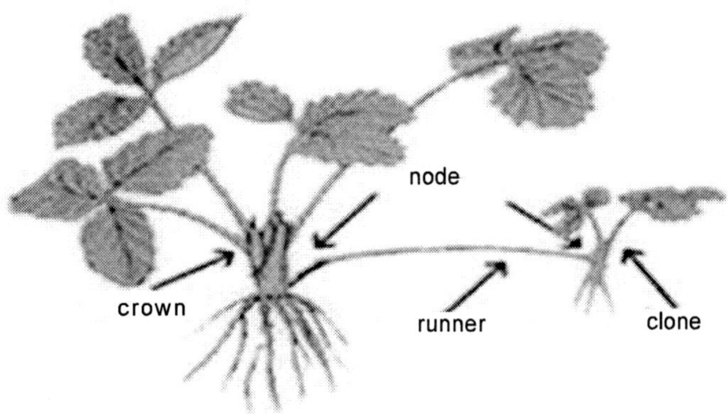

Strawberry runner

- All the plants with good root system should be utilised to set a new plantation. Given the best attention and care, a single plant usually produces 12 to 18 runners.

Planting

- The land for strawberry planting should be thoroughly prepared by deep ploughing followed by harrowing.
- Liberal quantities of organic manure should be incorporated in the soil before plating.

- Strawberry can be planted on flat beds, in the form of hill rows or matted rows, or it can be planted on raised beds.
- In irrigated areas, plantings on ridges are advised. In Mahabaleshwar, the usual practice is to plant on raised beds 4 x 3 meters or 4 x 4 meters.
- The planting distance should be 45 cm from plant to plant and 60 to 75 cm. from row to row.
- In the hills, Transplanting is done in March-April, September-October, but in the plains, the months of January-February may be utilised for this purpose.
- At Mahabaleshwar normally strawberry is planted during November-December.
- The plants should be set in the soil with their roots going straight down.
- The soil around the plant should be firmly packed to exclude air.
- The growing point of the plant should be just above the soil surface. During planting, the plants should not be allowed to dry out and should be irrigated immediately after planting.

Care of Young Plantation

- The roots of strawberry plants spread out close to the surface. Therefore, the soil
- should be well supplied with moisture, and hoeing should be done lightly and young plantation be kept weed free.

Special Horticultural Practices

- In cold climate the soil is covered with a mulch in winter to protect the roots from cold injury. The mulch keeps the fruits free from soil, reduces decay of fruits, conserves soil moisture, lowers soil temperature in hot weather, protects flowers from frost in mild climates and protects plants from freezing injury in cold climates.

- Several kinds of mulches are used, but the commonest one is straw mulch. The name strawberry has been derived from this fact.
- Black alkathine mulch is also used to cover the soil. It saves irrigation water, prevents the growth of weeds and keep the soil temperature high.

Irrigation

- Since strawberry is relatively shallow-rooted, it is susceptible to conditions of drought.
- Planting early in autumn allows the plants to make good vegetative growth before the onset of winter.
- However, in this case it is necessary to ensure that newly planted runners are irrigated frequently after planting; otherwise the mortality of the plants becomes high.
- During September and October, irrigation should be given twice a week if there is no rain.
- It may be reduced to weekly intervals during November.
- In December and January, irrigation may be given once every fortnight.
- Maybe when fruiting starts, the irrigation frequency ought to again be increased.
- At this stage frequent irrigation gives larger fruits.

Applica tion of Manures and Fertilisers

- Strawberry requires moderate amounts of nitrogen.
- Addition of organic matter to the soil, in the form of 50 tons of Farm Yard manure per hectare is highly desirable.
- It improves the water holding capacity of the soil and also gives better runner formation.

- Farm yard manure may be supplemented by chemical fertilizers to make up the total quantity of nitrogen from 84 to 112 kg per hectare, Phosphorus 56 to 84 kg per hectare, and Potash 56 to 112 kg per hectare.
- The Phosphatic fertilizer should be incorporated into the soil before plantings.
- The nitrogenous fertilizer be applied in Two doses (Three weeks after planting and again at the time of flowering) and potash at the time of flowering only.
- Application of adequate amounts of nitrogen gives higher yield of early berries.

Plant Protection

- Red spider mites and cutworms are important pests of strawberry.
- The mites can be controlled with 0.05 per cent Monocrotophos + 0.25 per cent wetablesulphur.

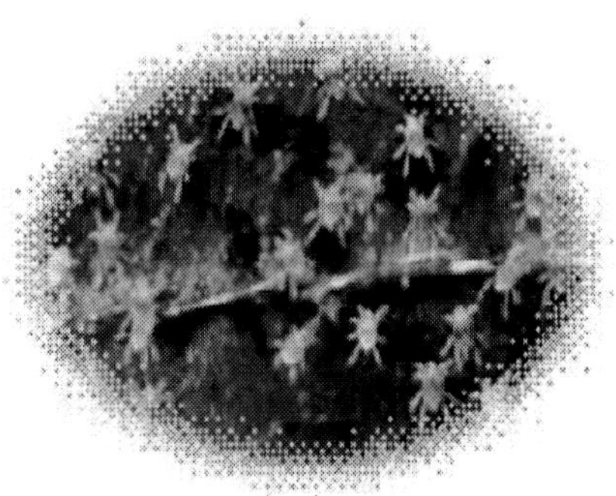

- The cut worms can be controlled by dusting the soil before planting with 5 per cent chloradane or Heptachlor dust at the rate of 50 kg per hectare and mixing it thoroughly in the soil by cultivator. · The two commonest diseases of strawberry are red stele, caused by the fungus *Phytophthora fragariae* and black root rot.

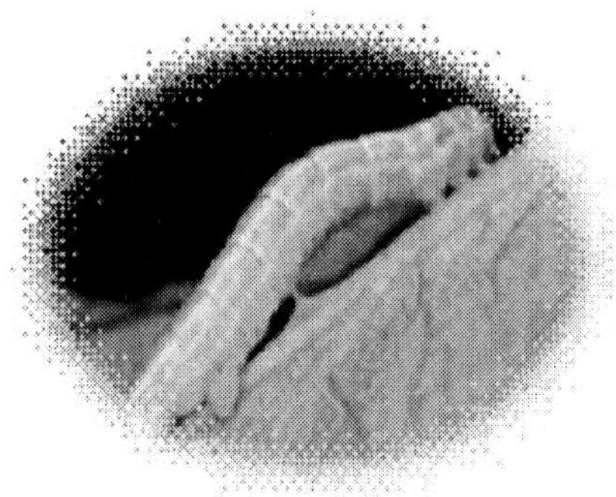

- The remedy for the former lies by growing resistant varieties like stelemaster and for the latter to maintain the vigour of the plants and rotate strawberry with other crops like legume vegetables (beans, peas etc).
- Strawberry also suffers from virus diseases known as yellow edge, crinkle and dwarf. Raising of strawberry nursery in the hills helps to check these.

- Strawberry also throws some chlorotic plants, which result from genetic segregation.
- These should not be confused with virus affected plants and should be rogued out.

Harvesting and Yields

- For local market the fruit should be harvested when fully ripe, but for transport to distant markets, it should be harvested when still firm and before colour has developed fully all over the fruit.
- Harvesting should be done preferably daily.
- Since fruit is highly perishable, it is packed in flat shallow containers of various types (cardboard, bamboo, paper trays etc.) with one or two layers of fruits.
- Harvesting should be done early in the morning in dry conditions. Washing the fruit bruises it and spoils its lustre.
- The yield varies according to season and locality.
- A yield of 20 to 25 tons per hectare is excellent, though yields upto 50 tons per hectare have been reported under ideal conditions.

Post-harvest Handling and

- Strawberries are highly perishable and hence a great deal of care in harvesting and handling as well as its marketing also requires to be organised carefully.

Marketing • Usually the fruit is picked in the early morning and sent to the market in the afternoon of the same day or is picked in the late afternoon, stored overnight in a cool place, and sent to market the following morning.

□□□

Chapter 18

Thrust Areas in Horticulture Sector

- Priorities on Horticulture and Plantation crops like tea, cashewnut, coconut, banana, ginger, turmeric, arecanut, etc. as these give higher revenue returns and encourage permanent cultivation thus having positive impact on the areas which hitherto had been under traditional *jhum* practices besides supporting food security.
- Economic transformation in production of horticulture crops with encouragement on growing of marketable crops and adoption of appropriate technology to ensure adequate production and economic returns.
- Strengthening of existing horticultural nurseries and encouraging of private nurseries for raising of quality planting materials in large numbers, setting up of tissue culture laboratories.
- Production of diseased-free certified and true-to-type seeds through establishment of seed farms in all the districts of the State to be supplemented with setting up of seed certification agency, introduction of seed village concepts in the pattern of registered growers for seed potato to be looked into as an alternative.
- Ensuring availability of irrigation facilities from minor irrigation projects, diesel pumps and shallow tube wells.
- Increasing the availability of proper horticultural machineries and tools through scheme for mechanization.
- Popularization of low–volume, high value items like dehydrated mushrooms, black pepper and other spices to be consciously attempted.
- Steps to be expedited to assess the viability of the two Government food processing units at Shillong and Dainadubi for smooth

operation on commercial basis to be made available through constant 'Revolving fund'.

- More emphasis on provision of poly-houses in both urban and rural setup for further increase of production of off-season vegetables, flowers and vegetable seeds in heavy rain-fed areas of the State.
- Schemes for tea cultivation which are currently limited to the three districts will also be extended to other districts in compact areas.
- Emphasis on new activities like floriculture, organic cultivation, etc. with special mention on cultivation of orchid which have export potential.
- Encouraging private entrepreneurs to set up small scale agro-based processing units near the horticulture cultivated areas for undertaking intermediate and preliminary processing of horticultural produce with suitable incentive from the State Government and package being offered by the National Horticulture Board (NHB). Department to also provide training support to entrepreneurs and farmers.
- Attempts to improve marketing facilities and marketing linkage and strengthening of minimum support price.
- Emphasis on the development of rural infrastructure for effective storage facilities and market infrastructure.
- Use of Information Technology through setting-up of computer network for imparting new technology and for providing marketing intelligence system.
- Implementation of existing developmental schemes in right earnest and formulation of schemes for employment generation that would provide ample scope for involving of un-employed to enable them to earn their livelihood. Schemes for tea cultivation which are presently limited only to the three districts of the State to be extended to other districts in compact areas.
- Popularisation and commercialisation of indigenous fruits and plants.
- Commercialisation of indigenous herbs and medicinal plants.

❑❑❑

Chapter 19

Turmeric

Botanical Name : *Curcuma longa*

Family Zingiberaceae

Uses of Turmeric

1. Used as a conditioner. 2. Used medicinally for external application and consumed as a stimulant.

Area and Production

India is the largest producer of turmeric with a annual production of 3.49 lakh tones (1991-92) and 4.35 lakh tones (96-97). The turmeric export-stands at 18.95 thousand, tones (92-93) and 25 M thousand tones (96-97). The major producing states are Tamil Nadu, Andhra Pradesh and Orissa.

Soil

1. Different kinds of soils such as sandy loam to clay loam or alluvial soils are suitable. Well-drained loamy soils are the best.
2. Soil should be rich in organic matter and uniform in texture.
3. Rich loamy soils having natural drainage and irrigation facilities are the best
4. Turmeric cannot withstand water stagnation or alkalinity.

Climate

1. Tropical crop, requiring warm and humid climate.
2. Temperature range-24°C to 28°C.
3. Growth ceases when, temperature falls below 20°C, and hence early-planted turmeric gives good yield.
4. Thrives well in localities with annual rainfall from 70 - 225 cm.
5. Can be grown at an altitude of 1200 m.

Cultivation of Turmeric

I) SeedMaterial

a) **Seed Selection**

1. Seed of turmeric consists of rhizomes.
2. Both mother and finger rhizomes are used.
3. The fingers are cut into pieces, each 4-5 cm long with 1-2 buds. Mother rhizomes are planted as such or split into two, each having one sound bud.
4. Mother rhizomes are preferred since they give 50% more yields than the finger rhizome and also give good growth.
5. Large sized, plumy and healthy mother rhizomes at least 100 g in weight should be used.

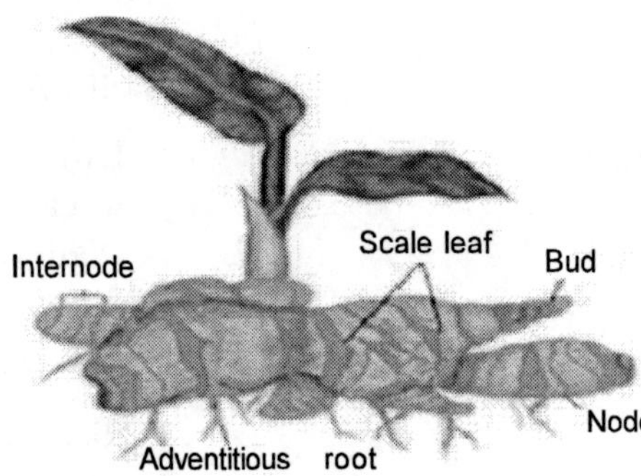

b) **Seed Rate**

1. Varies according to type of planting material, spacing and weight of rhizomes.

2. Mother rhizomes: 2000-2500 kg/ha.
3. Finger rhizomes: 1500-2000 kg/ha.
4. As an intercrop in fruit garden: 400 - 500 kg/ha.

c) **Seed Treatment:** Rhizomes are treated with 50 Lindane powder and Bavistine (2 g/1) by dipping for 15-20 minutes for prevention of diseases and rhizomes fly during early period.

II) Preparation of Land

1. Starts immediately after harvest of previous crop or with/onset of early rains in plains.
2. Land is ploughed 15-20 cm deep arid exposed to sun for one month. 2-3 crosswise harrowing are given.
3. Temporary ridges are opened to prevent soil erosion on slopy lands.

III) Systems of Planting

1. **Flat Beds**: Used under rainfed conditions where soils are light. Flat beds 1 m in width and-of suitable length varying according to the slope of land are prepared.

2. **Ridges and Furrows:** Under irrigated-conditions where the land is levelled or plain and soils are heavy, planting clone on ridges and furrows, opened at 75 cm distance and haying 3-3.5 m length. Broad ridges should having 90-100 cm width and 3-6 m length depending on the slope.

Planting of Ginger

Season of Planting: May to July depending on tract.

Method of Planting

(i) **On Flat Beds:** (25 x 2.5 cm) in each direction (ii) On ridges and furrows: 40-60 x 25 cm. Rhizomes are planted at 1/3rd height of ridge on broad ridge.

Manuring and Fertilization :

Turmeric is a heavy feeder crop. In Meghalaya on the basis of experiments conducted recommended fertilizer dose is; 25-31 ton FYM, 120 kg N, 50 kg P205 and 50 kg K20/ ha.

Time of Application

Sr. No.	Time	Manure (tones/ Ha)	N (kg/ ha)	P (kg/ ha)
1.	At preparatory tillage i.e. before last harrowing	30	-	-
2.	Basal application at planting	-	-	50
3.	One month form planting	-	60	-
4.	Two months from planting	-	60	-
	Total	30	120	50

Intercropping

Mixed crops like chilli, onion, brinjal and maize can be taken.

Aftercare

a) **Mulching:**

1. Done when-planted on raised beds
2. 2-3 mulching, are given
3. 1st immediately after planting
4. 2nd and 3rd at an interval of 40-50 days.

b) **Earthing up:**

1. To avoid exposure of developing underground rhizomes to sun due to soil erosion.
2. Done with light digging, 2 - 2.5 months after planting.

c) **Weeding:**

1. Pot is kept clean during first 4-6 weeks.
2. Depending on intensity of weeds, 5-6 weeding are given.

Irrigations

1. First irrigation is given before planting.
2. Second is given just after planting.
3. Subsequent irrigations are given at 7-10 days interval depending on soil.
4. A total of 20-25 irrigations are given, during the life period of the crop.

Harvesting

Starts from Feb. and continues till April.

2. Rhizomes are ready for harvest in 7-9 months after planting.
3. Longa type - 9 months, Amada type - 8 months and Aromatica type - 7 months.
4. State: Turmeric is harvested when leaves start yellowing and ultimately the stem dries down. The plants are-cut close to the ground.
5. The crop is irrigated lightly for easy digging.
6. Harvesting consists of digging of underground clumps of rhizomes with Pick axe or digging fork.
7. Fingers are separated from mother rhizomes.

Ipm/Insect Diseases Pest Management

1. **Leaf Blotch:** Small oval, rectangular or irregular brown spots appear on either side of leaves and soon become dirty brown. Yield is reduced. Control: Spray 0.2 % Diethene M-45.

2. **Leaf Spot:** Brown spots of different size appear on the upper surface of young leaves. The spots are whitish or grayish in the center. Subsequently, the whole leaf gets covered and leaves dry up. The rhizomes do not develop well.

Control: Spray 0.3 % Zineb or 1 % Bordeaux mixture.

3. **Rhizome Rot:** Leaves dry up starting from the margins. Collar region becomes soft, water soaked and the plants, collapse, Rhizome decay follows.

Control:

a) Treat the seed rhizomes with 0.3 % Diethane M-45 for 30 minutes before sowing and also before storage.

b) Under field conditions, drench the beds with 0.3 % Diethane M-45.

Insects/ Pests

1. **Shoot Borer**: The caterpillar bores into rhizomes and feeds on it.

   Control: Spray 0.1 % Malathion.

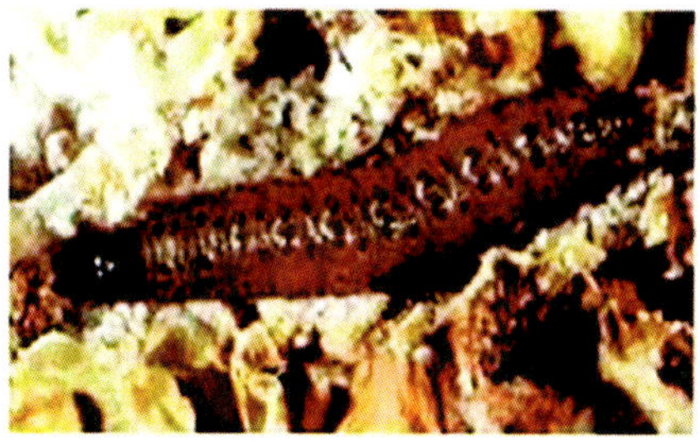

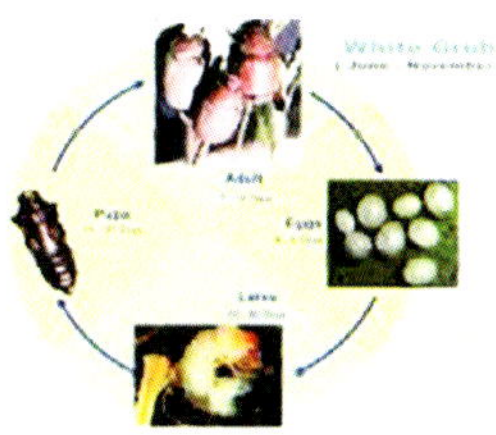

2. **Leaf Roller:** Caterpillar remains in leaf folds and feds within.

   Control: Spray 0.05 % Dimethoate.

3. **Scale Insects:** The scales, damage the rhizomes both in field and in storage

   Control: Dip seed rhizomes in 0.05 % Malathion or Dimethoate for 30 minutes before storage or planting.

Yield

250-300 q/ha for **fresh rhizomes**. Cured Produce: 20-25 % of fresh rhizome by weight.

Preservation of Seed

Rhizomes for seed are heaped in the shade of trees or in well ventilated sheds and covered with turmeric leaves. Sometimes the heap is plastered over with earth mixed with cow dung.

Quality of Finger

1. Should be free from chemichrome
2. Extraneous matter should be less than 2 %

Quality of Powder

1. Chemichrome test should be negative
2. Total ash less than 9 %
3. Moisture less than 13 %

Turmeric Oil

A volatile oil is obtained by steam distillation of ground turmeric after 8-10 hours. The oil is pale yellow to orange yellow in colour. The aroma of oil is due to tumerone and artumerous.

❑❑❑

Chapter 20

Ginger

Botanical Name *Zingiber officinalis*

Family Zingiberaceae

Origin South - East Asia, India

Area and Production of Ginger

- Ginger *is* important spice crop of India and accounts for **45 %** of the world's ginger production.
- Area under cultivation in India is about 63,000. ha with total production of about 2 lakh tones.
- The average productivity is about 3 tones/ha

Importance of Ginger

- Used as a spice on a large scale.

  o Used in the preparation of **medicines and confectionaries**.
- Flavor of ginger is due to **zingiferine**. **Pungency** of ginger is due to **zingirone.**

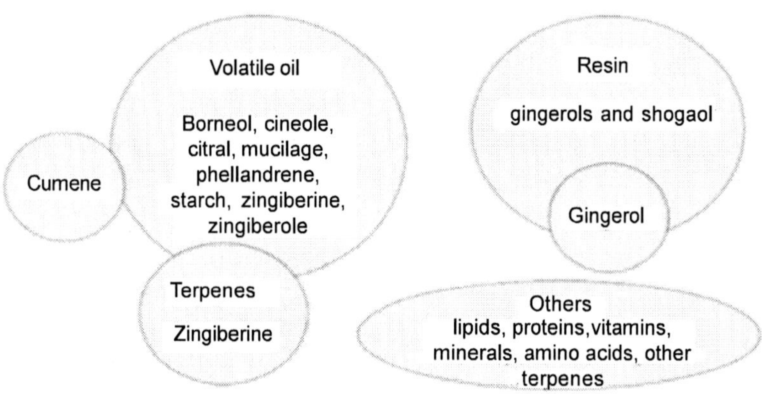

Soil Deep, well-drained, friable, loamy soil, rich in humus is ideal for ginger. Crop does not thrive well in alkaline soil. It is not desirable to grow ginger in the same field year after year.

Climate

- Tropical crop requires warm and humid climate. Grows well in areas with annual rainfall between 125 to 250 cm.Cultivated up to 1500 m above sea level. Cool and dry climate is best for rhizome development.
- Shade loving plant and requires ample moisture for normal growth.

Varieties Several cultivars are grown and are known by the particular area of the locality. The important cultivars are:

High Yielding Types : Maran, Karakkal, Rio de Janeiro, Mahim

Less Fiber Content : Jamaica, Bangkok, China Thingpuri

High Oleoresin: Emad Chemed, China, Karuppamadi, Rio de Jeneiro

High Dry Ginger Recovery : Karakkal, Nadia, Maran, Tura.

High volatile oil : Sleeva local, Narasapattam, Emad.Chemad.

Land Preparation

1. Land is ploughed 2 Times (crosswise) in summer (March - April) to a depth 15 to 22 cm.
2. Exposed clods are crushed with Norwegian harrow.
3. 3-4 crosswise harrowing are given to make soil loose and friable.
4. Temporary ridges are opened to prevent soil erosion on slopy lands.
5. 15 tons of FYM/ha is applied before last harrowing,
    - For raising rainfed crop, land is divided into raised beds of 1 m width and of convenient length varying from 3 - 6 m and 15 cm height with spacing of 30 cm between beds for drainage channel.

- On hill slopes, (the beds are formed along the contours. Ginger is planted in shallow pits in rows 25 cm apart and at a distance of 15-20 cm within the rows.
- · In case of irrigated crop, ridges 40-45 cm apart are opened. Planting is done in shallow pits on top of ridges at a distance of 22-30 cm.

Propagation

Ginger is propagated by using portions of mother rhizomes called assets. Each healthy set to be used for planting should be 2.5 to 5 cm long, weighing 20-25 g and having two or three buds each. The seed rhizomes should be treated with Dithane M-45 @ 3 g per liter of water for 30 minutes, drained and then used for planting.

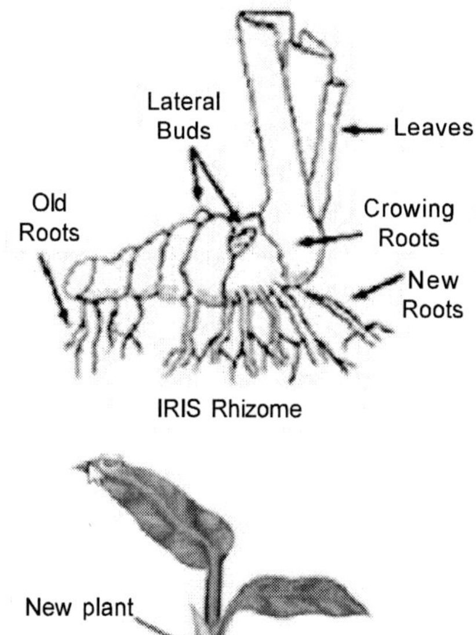

IRIS Rhizome

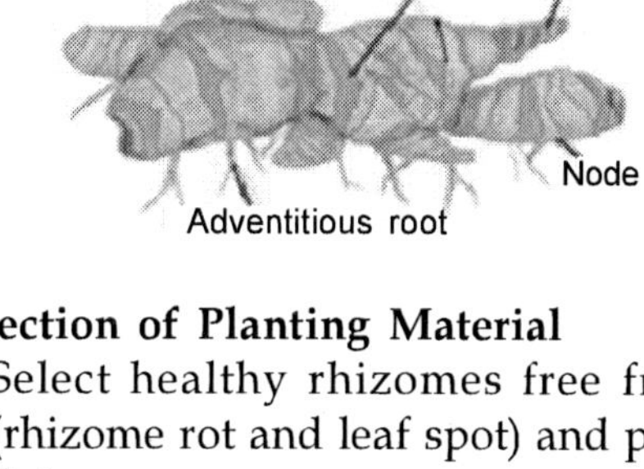

Seeds and Sowing:

a) **Selection of Planting Material**

1. Select healthy rhizomes free from diseases (rhizome rot and leaf spot) and pests (rhizome fly).

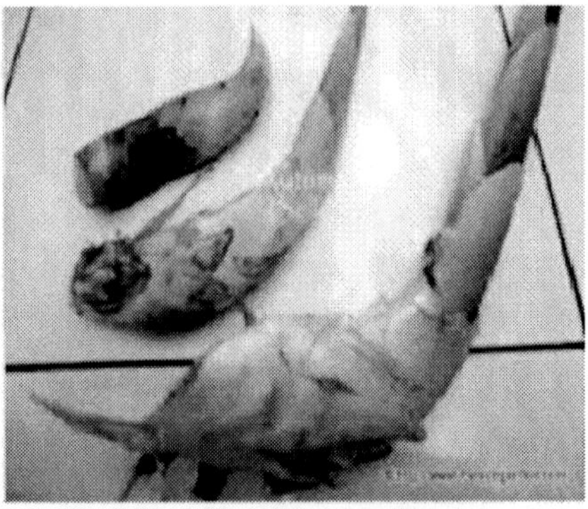

2. **Sprouted rhizomes** are broken into pieces keeping 2-3 sprouted eye buds on each rhizome.
3. Each piece should be 2.5-5 cm long and 20-25 g in weight.

b) **Time of Planting:**

1. **Ginger** can be planted from start of May up to middle of June.
2. 2nd forth night of May is found to be the best time for planting of ginger.

c) **Seed rate:** 1500 kg/ha mother rhizomes is the recommended seed rate.

d) **Preparation of Seed Material:**

- For planting on 1 ha area, 1500 kg rhizomes free from pests and diseases are selected 1 1/ 2month before planting.
- These rhizomes are treated with 25% agallol and Macrotophos for 30 minutes for control of rhizome fly.
- Treated rhizomes are stored in trenches prepared under shade. The trenches are bottoms-lined with dry leaves/straw. Rhizomes are spread on this arid are again covered with dried leaves/straw. This helps to control moisture loss and improves germination percentage.

Methods of Planting

a) **Ridges and furrows-** 75 cm apart for irrigated crop.

b) **Hot bed method**- 20-30 x 20-30 cm (25 x 22.5 cm)

c) **Raised beds -** 30 x 30 cm.

d) **New improved Technique**- Planting is done by broad ridge method (75 cm broad, 20-30 cm high and 30 cm apart).

Pre-sowing irrigation is given and planting is done at 'Wapsa' condition. During planting, the tip of eye buds should face upwards. The rhizomes are planted at a depth of 5 cm.

Manures and Fertilizers

Sr. No.	Time of application	F.Y.M. (ton/ ha)	N (kg/ ha)	P2O5 Jkg/ *ha*	K20 (kg/ ha)
1.	Preparatory tillage	15	-	-	-
2.	At planting	15	60	50	50
3.	45 days after planting	-	50	-	-
4.	120 days after planting	-	40	-	-
	Total	30	150	50	50

Irrigation

1. First light irrigation is given immediately after planting.
2. Subsequent irrigations are given at 10 days intervals.
3. Total 16-18 irrigations
4. Water requirement of crop is 90-100 ha cm.

Inter Culturing

a) **Mulching**: It is followed in rain fed crop to prevent weed growth, for protection from sun, preventing evaporation losses, to maintain soil temperature for protection from heavy rains and for consequent enhancement of organic matter.

b) **Shading:** Seeds of cluster bean, pigeon pea or castor are sown in irrigation channels on the comer of raised beds for shade.

c) **Weeding:**

1. The plot is kept clean by hand weeding during first 4-6 weeks.
2. Depending upon intensity of weeds, 5-6 weeding are given to have better yield.

d) **Earthing-up:**

The soil around the plants is worked with the help of *khurpi* in the first week of September. It helps to break the fibrous roots and thereby supports new growth. The soil near the rhizomes becomes loose and friable and helps in proper development of rhizomes.

IPM / Insects and Diseases Management of Ginger

Diseases

Diseases of Ginger

Bacterial wilt : *Ralstonia solanacearum*

Symptoms: It is the most serious disease and the symptoms can be noticed form July-August. The leaf margins of the affected plant turn bronze and curl backward. The whole plants wilt and die. The base of the infected pseudostem and the rhizome emit foul smell. When the suspected pseudostem is cut and immersed in a glass of clean water, milky exudates will ooze out from the cut end. Typical symptom is the wilting observed during afternoon in young seedlings.

Management: Seed contamination is the major source of infection. Hence, procure only healthy rhizome from disease free area. Treat the seed with Streptocyclin (20g/100 litre water). Remove the affected clumps and drench the soil with copper oxychloride 0.2%.

Soft rot: *Pythium aphanidrematum*

Symptoms: It is a serious seed as well as soil borne disease and the symptoms can be seen from July. Yellowing of leaves appear first on the lower leaves and proceeds to upper leaves. Roots arising form the affected rhizome become rotten and show brown discoloration of the rhizome tissue. Sometimes the pseudostem comes off easily with a gentle pull. The rotten parts attract other fungi, bacteria and insects particularly the rhizome fly. During the rainy season, this disease spreads very fast from infected field to healthy field.

Management : Avoid water logging. At the time of sowing, treat the rhizome with Bordeaux mixture (1%) and again with Trichoderma @8-10-gm/litre water.

Remove the badly affected plants and drench around the infected plants, after slightly removing of soil with Bordeaux mixture (1%) or copper oxychloride @ 2g/ 1 liters of water.

Dry rot : ***Fusarium and Pratylenchus complex***

Symptoms : It is a fungus-nematode complex disease. In contrast to rhizome rot, dry rot appears in field in small patches and spreads slowly. The affected plants appear stunted and exhibit varying degree of foliar yellowing. Older leaves dry up first followed by younger ones. In advanced stage the rhizome, when cut open, show a brownish ring and is mainly restricted to cortical region. The pseudo stem of the dry rot affected plants does not come off with a gentle pull in contrast to soft rot. The affected rhizomes are shrunken, dry and are not marketable.

Management : Soil application of mustard oil cake at the rate of 40 kg/ha before sowing in furrows can check the nematode problem. Hot water treatment (51OC for 10 min) followed by seed treatment with Bordeaux mixture (1%) effectively checks the problem.

Leaf spot / blight : ***Phyllostricta zingiberi***

Symptoms : Small spindle to oval spots appear on younger leaves. The spots have white papery centers and dark brown margins surrounded by yellowish halos. The spot later increase in size and coalesce to form larger spots which eventually decrease the photosynthetic area. In the case of severe infection the entire leaves dry up.

Management : Spray Bordeaux mixture (1%) 3-4 times at 15 days interval with the initiation of the disease. Good control is achieved by growing the crop under partial shade.

Insects pests

White Grub or: *Holotrichia* spp.

It is a sporadic pest, sometimes causes serious damage. The grub feeds on the roots and newly formed rhizomes. The infestation is generally more during August-September. The adult beetles, after emergence from pupae settle on the Ficus or other trees in congregation which can be collected and destroyed. The entomophagous fungus Metarrhiziuma-nisophilae can be mixed with fine cow dung and then applied in the field to control the grubs. In endemic areas opt for soil application of neem cake @ 40 kg/ha before sowing.

Shoot borer :*Conogethes punctiferalis*

The larvae bore the tender pseudostem and reach the central portion by feeding on the internal tissues, thus resulting in yellowing and drying of shoots. Infestation may occur from June to October. Spray Nimbicidine (2-5ml/l) or Beauveriabas-siana@ 2-5ml/l

Shoot boring weevil: ***Prodiocteshae maticus***

The grubs bore into the pseudostem and cause dead hearts. Remove alternate host plants such as wild turmeric and cardamom. The congregating adult beetles can be collected and destroyed. Spray Nimbicidine @ 2-5ml/l or Carbofuran 3G granules @ 30 kg/ha immediately after mother rhizome extraction. In the case of severe infestation, spray Endosulfan @ 0.07%.

Integrated Pest and Disease Management for Ginger

- Field hygiene is more important to manage the pests and diseases. Avoid water stagnation, provide adequate drainage, remove weeds periodically, apply only well rotted FYM compost @ 25 t/ha and thoroughly incorporate it in the soil, apply dolomite @ 2 t/ha before sowing to increase soil pH, sow ginger in raised beds of at least 25-30 cm height and provide mulching with leaves and twigs of chilaune (Schimawallichii) or Banmara (Eupatorium sp) or utis (Alnusnepalensis) or mustard oil cake @ 5 to 10 t/ha and follow crop rotation of 2 to 4 years depending on the incidence and severity of the diseases. Soil application of Biocontrol agents like T. harzianum and P. fluorescence during planting time @ 2-5% gives effective control of the diseases.

- Use good quality rhizome for sowing. Procure disease free seeds from disease free area.
- Before sowing, treat the rhizome in hot water (51oC for 10 min) and again in solution of Bordeaux mixture 1% for 15 min. Add Streptocyclin (20g/ 100 l water) if bacterial wilt is also a problem. Dry the rhizome in shade and then sow. If cut rhizome are to be planted, they should be treated after cutting.
- Treat rhizome with bio-inoculant Pseudomonas fluorescens & Trichoderma harzianum followed by soil application 60 days after planting to reduce rhizome rot.
- Once the diseases is spotted in the field, remove the affected clumps and drench the soil with Bordeaux mixture 1% at 15 days interval.
- Diseased plants should be identified while the crop is in field. Rhizomes from such plants should not be selected for the seed purpose
- Mechanical collection and destruction of grubs, weevils, larvae and adult beetle periodically will reduce the incidence of insect pests. If white grub is predominant, apply Nimbicidine @ 2-5 ml/L.

Lacewing

Ladybird beetle

Spider

Predatory thrips

Praying mantis

Hover fly

- It has been observed that diseases spread fast after mother rhizome extraction. Hence, drench the soil with fungicide immediately after mau extraction and again this practice has to be followed keeping into consideration the cost - benefit ratio of mau extraction and disease incidence.

Harvesting:

1. For green ginger, harvesting is done 210-215 days after planting.
2. For curing purpose, harvesting is done 245-260 days after planting when the leaves start yellowing and pseudo stem begins to dry.
3. The rhizomes are lifted either with digging fork or using *kudali*and are cleaned-off the roots and adhering soil particles.

Yield of Cured Ginger 15-20 % of fresh produce.

Processed products of ginger

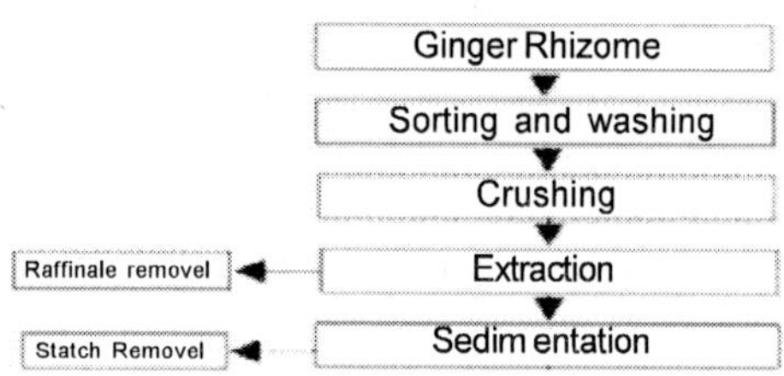

Glossary of Horticultural Terms

Air drainage. The flow of cold air down a hill. Plant crops that are sensitive to late spring and early fall frost on slopes, preferably south-facing, so that cold air will drain downhill and settle in the low spots below the planted area.

Air layering. A propagation method. A cut is made three-quartets of the way through the stem at an angle of 45 degrees. A toothpick is inserted to keep the cut from sealing itself closed. The cut is wrapped in sphagnum moss that has been soaked in water and wrung out. This is then wrapped in plastic and tied at the top and bottom with ties.

Alternate leaf arrangement. Leaves are arranged at one leaf per node on different sides along the stem.

Annuals. Plants that complete their life cycles in one year or less.

Anther. The part of the flower (on the upper part of the stamen) in which pollen is produced.

Anthracnose. A type of leaf or fruit spot disease caused by a fungus.

Arms. In a grape plant, all main branches two years and older.

Asexual reproduction. Duplicating a plant from any cell, tissue or organ of that plant.

Axillary buds. Buds that are borne laterally on the stem in the axils of the leaves.

Biennials. Plants that require 2 years, or 2 growing seasons with a dormant period in between, to complete their life cycles.

Biological control. Reducing pests by utilizing other organisms. For example, controlling gypsy moths by using Bacillus thuringensis, a disease that affects the *gypsy* moth and is readily available in garden centers.

Blade. The flattened, green portion of the leaf.

Bolting. Going to seed. Plants will produce a flower stalk and seeds, and then die before the end of the season. Lettuce and other crops will bolt during very hot temperatures. Unless you are growing your crop for the seeds, you want to delay bolting.

Botrytis blight. A gray, felt-type mold that covers parts of a plant and causes stunting, dieback, and distorted growth.

Branch. The part of the tree that arises from the trunk and supports shoots, twigs and leaves. Budding. A type of grafting in which a non-fruitful (vegetative) scion is placed in a stock plant.

Buds. An undeveloped stem, consisting of a tiny bundle of cells, from which leaves, lateral buds, flower parts or all three will arise.

Bulb. A budlike structure that consists of a small stem with closely crowded fleshy or papery leaves or leaf bases.

Bulbil. An aerial bulblet.

Bulblet. Miniature bulbs that develop from meristems in the axils of scaly bulbs.

Canes. Previous season's growth from the arms or trunk of a grape plant; or, the supporting part of the raspberry plant, similar in function to the trunk of a tree.

Capsule. Simple, dry, dehiscent (see definition of dehiscent) fruit with two or more locules (the cavities of the ovary of the pistil of a flower) which split in various ways.

Catkin. A type of spike arrangement of flowers on the floral axis that has unisexual flowers made up of sepals and petals.

Certified seed. The progeny of registered seed stock. It is the final stage in the expansion program and is certified with a metal seal and blue tag.

Chilling injury. Damage to certain horticultural products, such as banana, papaya, cucumber and sweet potato, which results from exposure to cold but above-freezing temperatures.

Chilling requirement. A cold period required by certain plants and plant parts to break physiological dormancy or rest. The chilling requirement is expressed in terms of the required number of hours at 7 degrees centigrade or less.

Cold storage. A type of insulated storage utilizing mechanical refrigeration to maintain a stable, cold temperature for long-term storage.

Commercial floriculture. Area of horticulture that includes the commercial production and distribution of cut flowers, flowering pot plants, foliage plants and bedding plants.

Complete fertilizer. A fertilizer that contains nitrogen, phosphorus, and potassium.

Complete flower. Flower composed of a short axis or receptacle from which arise four sets of floral parts-sepals, petals, stamens, and pistils.

Compound leaf. A leaf with the blade divided into several leaflets or sections.

Compound umbel. A type of flower structure in which a series of simple umbels (see definition of umbel) arise from the same point on the main axis.

Corms. Short, fleshy, underground stems with few nodes and very short spaces between nodes. Gladiolus and crocus are cormous plants.

Corymb. Type of raceme flower cluster in which the pedicels of the lower flowers are longer than the pedicels of the upper flowers, resulting in a flattopped flower cluster.

Cross-pollination. The process in which pollen is transferred from an anther (the upper part of the stamen in which pollen is produced) of one flower to the stigma (the pollen-receiving site of the pistil) of a second flower of a different cultivar.

Cultivar. A plant derived from a cultivated variety that has originated and persisted under cultivation, not necessarily referable to a botanical species, and of botanical or horticultural importance, requiring a name.

Cuticle. The waxy covering on leaves or fruit, which protects the tissue against excess moisture loss.

Cuttings. Detached vegetative plant parts which when placed under conditions favorable for regeneration will develop into a complete plant with characteristics identical to the parent plant

Cyme. A broad, more-or-less flat-topped determinate (having a fixed number) flower structure in which the central flowers bloom first.

Day-neutral plant. A plant that will flower under any daylength.

Deciduous. Trees or plants that shed their leaves every year.

Dehiscent. Type of dry fruit in which the carpel (a simple pistil) splits along definite seams at maturity. Determinate growth. Limited growth.

Dioecious. Species that produce male and female flowers on separate plants.

Disbudding. The removal of vegetative or flower buds.

Diseased plant. A plant that is abnormal because of a disease-causing organism or virus.

Dormancy. A state of suspended growth or the lack of outwardly visible activity caused by environmental or internal factors.

Drip irrigation. The application of small quantities of water directly to the root zone through various types of delivery systems on a daily basis.

Drupe. Type of fruit with a thin outer skinlike region, a center portion that is thick and fleshy, and an inner area that is hard and stony.

Endocarp. The inner area of the fruit wall. Entomology. The study of insects.

Epidermis. The outer layer of cells of all parts of a young plant and of some parts of older plants, such as leaves and fruits. These cells are usually covered with a waxy substance, called the cuticle, that minimizes water loss.

Exocarp. The outer skinlike region of the fruit wall.

Fallow. A system in which land is left without a crop for an extended period to accumulate moisture and restore nutrients.

Fleshy fruits. Classification of fruits that includes the berry, drupe, and pome. They have a fruit wall that is soft and fleshy at maturity.

Floriculture. The study of growing, marketing, and arranging flowers and foliage plants.

Flower. A shoot of determinate (limited in number) growth with modified leaves that is supported by a short stem; the structure involved in the reproductive processes of plants that bear enclosed seeds in their fruits.

Forcing. The manipulation of environmental factors which makes it possible to produce a marketable pot plant or cut flower out of season.

Foundation seed. Seed stock handled to most nearly maintain specific genetic identity and purity under supervised or approved production methods certified by the agency.

Frost pocket. A depression in the ground into which cold air drains but from which it cannot escape, thus causing it to be an area very subject to freeze injury.

Fruit. An expanded and ripened ovary with attached and subtending reproductive structures. Genes. Basic units of hereditary material that dictate the characteristics of individuals.

Germination. The initiation of active growth by the embryo, resulting in the rupture of seed coverings and the emergence of a new seedling plant capable of independent existence.

Grafting. The joining of two separate structures, such as a root and stem or two stems, so that by tissue regeneration they form a union and grow as one plant

Growing season. The period from the last spring freeze until the first freeze in the fall. In the United States this ranges from about 100 to 365 days.

Hardening. The result of many changes that occur in a plant as it develops resistance to adverse conditions, especially cold.

Hardening off. The treatment of tender plants to enable them to survive a more adverse environment. Treatments involve withholding nutrients, lowering temperatures, allowing temporary wilting, and other methods to slow growth rate.

Hardpan. A layer in the soil that cannot be penetrated and which restricts root penetration as well as movement of air and water.

Hardwood cuttings. Cuttings made from woody deciduous species and narrow-leaved evergreen species, such as grape and hemlocks.

Heading back. Type of pruning cut where the end portion of the shoot is removed but the portion at the base is not.

Herbaceous cuttings. Cuttings made from herbaceous plants (see definition below), such as chrysanthemums, coleus, and geraniums.

Herbaceous perennials. Plants with soft, succulent stems whose tops are killed back by frost in many temperate and colder climates, but whose roots and crowns remain alive and send out top growth when favorable growing conditions return.

Herbarium. A collection of plant specimens that have been taxonomically classified, pressed, dried, and mounted on a sheet of herbarium paper.

Herbicide. A material that will kill plants. Herbicides may kill essentially all plants or be quite selective in their activity.

High-pressure sprayer. A sprayer using a high-pressure pump to force the spray through nozzles for both creating a fine spray and delivery to the plant.

Horticulture. The intensive cultivation of plants.

Humus. A dark-colored organic material that is the product of organic matter breakdown by both microorganisms and chemical reactions.

Hybrid. In its simplest form, a first-generation cross between two genetically diverse parents. Imperfect flower. Flower lacking either the stamen or the pistil.

Incomplete flower. Flower lacking one or more of the fours sets of floral parts.

Indehiscent. Type of dry fruit in which the fruit wall does not split at any certain point or seam at maturity. Indeterminate growth. Growth that is potentially limitless.

Inflorescence. The arrangement of the flowers on the floral axis-a flower cluster.

Interstock. Intermediate stem piece that is grafted between the upper and lower pars of the graft.

Landscape design. The profession concerned with the planning and planting of outdoor space to secure the most desirable relationship between land forms, architecture, and plans to best meet human needs for function and beauty.

layering. A vegetative method of propagation that produces new individuals by producing adventitious roots (roots growing from unusual places) before the new plant is separated from the parent plant.

Leaching. The downward movement of nutrients or salts through the soil profile in soil water. Leaching accounts for nutrient losses but can also be beneficial in ridding a soil of excess salts.

Leaf arrangement. The arrangement of leaves on the stem.

Leaf-bud cuttings. Cuttings consisting of a leaf blade, leaf stalk, and a short piece of the stem. Leaf cuttings. Entire leaves with or without the leaf stalks.

Leaflets. Individual parts of a compound leaf.

Leaves. Vegetative plant parts that are lateral outgrowths of stems that have developed special structures for photosynthesis.

Legume. A simple, dry, dehiscent (see definition of dehiscent) fruit with one locale that splits along two seams.

Lime. Ground limestone that is used to raise soil pH.

Loam. A type of soil consisting of sand, clay, and organic matter. Locales. The cavities of the ovary of the pistil of a flower.

Long-day (short-night) plant. A plant that requires a day longer than its critical daylength to flower.

Malling-Merton series. A group of apple rootstocks that were the result of the crossing the Malling series with the 'Northern Spy' to add resistance to woolly aphids.

Malling series. A group of apple rootstocks that originated at the East Malling Research Station in England.

Market gardening. The growing of an assortment of vegetables for local or roadside markets. Mesocarp. The center portion of the fruit wall.

Micronutrient. An essential nutrient that is needed in small amounts; also called a trace or minor element. Monoecious. Plants that produce male and female flowers on the same plant.

Mound or stool layering. A layering procedure in which the new shoots develop in spring and soil is mounded around their bases, excluding light and enhancing root formation.

Mulch. A material applied to the surface of a soil for a variety of purposes such as conservation of moisture, stabilization of soil temperature, and suppression of weed growth.

Nursery grower. A person who produces or distributes ornamental plants.**Nut.** An indehiscent dry fruit similar to the achene except that the fruit wall is hard throughout. Offsets. Miniature bulbs grown to full size.

Offshoots. Short, horizontal stems that occur in whorls or near whorls at the crown of stems.

Olericulture. The study of vegetable production.

Opposite leaves. Leaves arising from opposite sides of the same node.

Organic matter. Materials rich in carbon of either plant or animal origin, which exist in all stages of decomposition of soils.

Ovary. Part of the pistil that contains one or many small bodies known as ovules. Ovule. The immature seed in the ovary.

Palmate leaf. A compound leaf with all the leaflets arising from one point at the end of the leaf stalk

Palmately veined leaves. Leaves with netted veins having several large veins radiating into the blade from the leaf stalk at the level where the leaf stalk and blade join.

Panicle. Type of flower cluster with either a cluster of racemes or corymbs (see definitions), distinctly branched.

Parallel venation. Leaves with large veins that are essentially parallel to one another and are not connected by lateral veins.

Peduncle. The short stem of the flower cluster.

Pepos. Berries that have a hard rind around the fruit.

Perennials. Plants that do not die after flowering but live from year to year.

Perfect flower. A flower that has both a pistil (or pistils) and stamens but may lack sepals, petals, or both. Perianth. Structure consisting of the sepals and petals.

Pericarp. The fruit wall, consisting of three distinct layers; the exocarp, the mesocarp, and the endocarp.

Pest management. The control of a pest or group of pests by a broad range of techniques from biological control to pesticides. The goal is to keep damage below economic levels without completely eliminating the pest.

Petals. Structures collectively making the corolla, which protect the inner reproductive structures and often attract insects by either their color or their nectar and thus facilitate pollination.

Petiole. The leaf stalk.

pH. A measure of acidity or alkalinity, expressed as the negative log of the hydrogen ion concentration. A pH of 7 is neutral; less than 7 is acidic more than 7 is basic

Pinching. Breaking off the uppermost growing point, thus allowing the willary buds to start to grow. This promotes outward growth rather than upward growth.

Pinnate leaf. A compound leaf with the leaflets arranged along both sides of the midrib.

Pinnately veined leaves. Leaves with netted veins where secondary veins extend laterally from a single midrib.

Pistil. The female reproductive organ, consisting of the stigma, style, and ovary. Plant growth. A permanent increase in volume, dry weight, or both

Plant pathogen. A microorganism that causes a plant disease.

Plant propagation. Increase in numbers or perpetuation of a species by reproduction.

Pome. Type of fruit in which the portions produced by the pericarp are enclosed within fleshy parts that are derived from parts of the flower other than the ovary. An apple is a pome.

Pomology. The study of fruit production.

Pot-bound. Having restricted root growth and a circular pattern of root growth caused by a too-small container.

Pruning. Removal of plant parts such as buds, developed shoots, and roots to maintain a desirable form by controlling the direction and amount of growth.

Raceme. Type of flower cluster in which stalked flowers are on pedicels approximately equal in length on a single floral axis.

Rest. A state of suspended growth or outwardly visible activity due to internal physiological factors. Also referred to as physiological dormancy.

Rhizome. Horizontal stems that grow underground. They serve as storage organs and a means of vegetative reproduction.

Root. Vegetative plant part that anchors the plant, absorbs water and minerals in solution, and often stores food.

Root cuttings. Cuttings 5 to 15 cm long made from root sections in the fall or winter.

Runner. A slender stolon (see definition) with elongated internodes. These root at the nodes that touch the ground. Example: strawberries.

Rust. A type of fungal disease that requires two hosts. For example, White Pine Blister Rust alternates between white pine and Ribes sp. Gives a rusty appearance to the plant.

Samara. An indehiscent dry fruit with either one or two seeds, in which pericarp bears a flattened wing-like outgrowth. The "helicopters" of the maple trees are double-samaras.

Scales. Modified leaves that protect structures.

Scaling. A type of propagation that separates bulb scales from the mother bulb to induce the formation of bulblets.

Scarification. The chemical or physical treatment given to some seeds to break or weaken the seed coat sufficiently for germination to occur.

Scion (cion). The upper part of the union of a graft.

Seed. Plant embryo with associated stored food encased in a protective seed coat.

Self-pollination. The process by which pollen is transferred from the pollen producing section of the plant to the pollen receiving part of the plant of the same flower or another flower of the same cultivar.

Semihardwood cuttings. Cuttings made from woody, broad-leaved evergreen species such as ligustrum and holly.

Sepals. Structures that usually form the outermost whorl of the flower; collectively, the calyx.

Separation. The use of bulbs and corms in propagation that utilizes the naturally detachable parts.

Serrate leaf. A leaf with serrations or "teeth" along the edge of the leaf.

Sessile. Without a petiole (as in some leaves), or without a pedicel (as in some flowers and fruits).

Sexual reproduction. The reproduction of plants through a sexual process involving cell division.

Shoot. Stem, 1 year old or less, that possesses leaves.

Simple bud. Bud containing either leaf or flower primordia, but not both.

Simple layering. A method similar to tip layering, except that the stem behind the end of the branch is covered with soil and the tip remains above ground.

Simple leaves. Leaf blades consisting of one unit.

Sinus. The indentation of a leaf. Sinuses alternate with lobes.

Slow-release fertilizer. A fertilizer that is made by coating the particles with a wax or other insoluble or very slowly soluble material to provide a predictable, slow release of the encapsulated materials.

Softwood cuttings. Cuttings taken from soft, succulent, new spring growth of deciduous or evergreen species of woody plants.

Soil. The outer, weather layer of the earth's crust that has the potential to support plant life. Soil is made up of inorganic particles, organic matter, microorganisms, water, and air.

Soil management. The practices used in treating a soil, which may include various types of tillage and production systems.

Soil structure. The arrangement of individual soil particles.

Spike. Flower cluster of an indeterminate (unlimited number) raceme of sessile flowers attached to the floral axis with the oldest flowers at the base.

Spikelet. Diminutive type of spike inflorescence found in grasses.

Spines. Sharp-pointed woody structures, usually modified from a leaf or part of a leaf.

Stamen. Part of the flower consisting of the anther, in which pollen is produced, and a slender filament that holds the anther in a position favorable for pollen dispersal. Male reproductive organ.

Staminate flower. Flower in which only the stamens (male reproductive parts) are present; there are no pistils (female reproductive parts).

Stigma. The pollen-receiving site of the pistil.

Stipules. Leaflike appendages often found on either side of the base of the petiole; they may subtend the leaf, but they are not actual parts of the leaf.

Stock (rootstock). The lower part of a graft.

Stolon. An aboveground stem that reclines or becomes prostrate and may form roots at the nodes that may come into contact with the ground.

Stratification. The storing of seeds at low temperatures under moist conditions to break physiological dormancy or rest.

Succulent leaves. Water storage structures characteristic of plants in and and semiarid regions.

Taproot. Primary root that persists and maintains its dominance.

Temperate fruit. A fruit plant that requires a cool period and is deciduous. Examples are apple, pear, and peach.

Tendrils. Slender, twining leaf modifications used for support, as in the terminal leaflets of the grape.

Terminal buds. Large and vigorous buds at the tips of stems, responsible for terminal growth.

Thorn. A short, sharp-pointed branch.

Tip layering. Layering in which rooting takes place near the tip of the current season's shoot, which naturally falls to the ground.

Tissue culture. The growing of masses of unorganized cells on agar or in liquid suspension. Useful for the rapid asexual multiplication of plants.

Topworking. Grafting procedure by which branches of trees are changed to a more desirable cultivar.

Trickle irrigation. The application of small quantities of water directly to the root zone through various types of delivery systems on a daily basis.

True leaves. The first leaves to emerge after the germination of a seed.

Tubers. Enlarged underground stems serving as storage organs of starch or related materials. Potatoes are tubers.

Twig. Stem, 1 year old or less, without leaves.

Umbel. Type of flower cluster in which the pedicels arise from a common point and are about equal in length.

Variegated. Leaves with green and white striped pattern instead of the typical all-green look.

Vegetable. The edible portion of an herbaceous garden plant.

Weed. A plant growing where it is not wanted.

Wet feet. A condition where plants are exposed to excess soil moisture caused by flooding or a high water table.

Whorled arrangement. Three or more leaves at a node.

Wilt. Loss of rigidity and drooping of plant parts caused by many different plant diseases. Generally, it involves an interference with water conduction inside the stems.

Woody perennials. Plants with woody fiber.

Yellows. A type of disease, common among some flowers, that is caused by a virus or mycoplasma-tike organism. This disease causes stunting and a yellowish appearance.

□□□

References

Borges, J.G, Nordström, E.-M. Garcia Gonzalo, J. Hujala, T. Trasobares, A. (eds). (2014). " Computer-based tools for supporting forest management. The experience and the expertise world-wide. Dept of Forest Resource Management, Swedish University of Agricultural Sciences. Umeå. Sweden.

Community of Practice Forest Management Decision Support Systems, http://www.forestdss.org/

Delic, K.A., Douillet, L. and Dayal, U. (2001). "Towards an architecture for real-time decision support systems:challenges and solutions.

Diasio, S., Agell, N. (2009). "The evolution of expertise in decision support technologies: A challenge for organizations," CSCWD, pp. 692–697, 13th International Conference on Computer Supported Cooperative Work in Design, 2009. http://www.computer.org/portal/web/csdl/doi/10.1109/CSCWD.2009.4968139

DSSAT4 (pdf)

Efraim Turban; Jay E. Aronson; Ting-Peng Liang (2008). Decision Support Systems and Intelligent Systems. p. 574.

Ender, Gabriela; E-Book (2005–2011) about the OpenSpace-Online Real-Time Methodology: Knowledge-sharing, problem solving, results-oriented group dialogs about topics that matter with extensive conference documentation in real-time. Download http://www.openspace-online.com/OpenSpace-Online_eBook_en.pdf

F. Burstein; C. W. Holsapple (2008). Handbook on Decision Support Systems. Berlin: Springer Verlag.

Gachet, A. (2004). Building Model-Driven Decision Support Systems with Dicodess. Zurich, VDF.

Gadomski, A.M. *et al.* (2001). "An Approach to the Intelligent Decision Advisor (IDA) for Emergency Managers", *Int. J. Risk Assessment and Management,* Vol. 2, Nos. 3/4.

Haag, Cummings, J141K1McCubbrey, Pinsonneault, Donovan (2000). Management InformatM1H1ion Systems: For The Information Age. McGraw-Hill Ryerson Limited: 136-140. ISBN 0-07-281947-2

Hackathorn, R. D., and P. G. W. Keen. (1981, September). "Organizational Strategies for Personal Computing in Decision Support Systems." *MIS Quarterly*, Vol. 5, No. 3.

Haettenschwiler, P. (1999). Neues anwenderfreundliches Konzept der Entscheidungsunterstützung. Gutes Entscheiden in Wirtschaft, Politik und Gesellschaft. Zurich, vdf Hochschulverlag AG: 189-208.

Henk G. Sol *et al.* (1987). Expert systems and artificial intelligence in decision support systems: proceedings of the Second Mini Euroconference, Lunteren, The Netherlands, 17–20 November 1985. Springer, 1987. ISBN 90-277-2437-7. p.1-2.

Holsapple, C.W., and A. B. Whinston. (1996). Decision Support Systems: A Knowledge-Based Approach. St. Paul: West Publishing. ISBN 0-324-03578-0

Jiménez, Antonio; Ríos-Insua, Sixto; Mateos, Alfonso. Computers & Operations Research.

Jintrawet, Attachai (1995). A Decision Support System for Rapid Assessment of Lowland Rice-based Cropping Alternatives in Thailand. *Agricultural Systems* 47: 245-258.

Keen, P. G. W. (1978). Decision support systems: an organizational perspective. Reading, Mass., Addison-Wesley Pub. Co. ISBN 0-201-03667-3

Keen, Peter; (1980). "Decision support systems : A Research perspective." Cambridge, Mass. : Center for Information Systems Research, Alfred P. Sloan School of Management.http://hdl.handle.net/1721.1/47172

Marakas, G.M. (1999). Decision support systems in the twenty-first century. Upper Saddle River, N.J., Prentice Hall.

Matsatsinis, N.F. and Y. Siskos (2002). Intelligent support systems for marketing decisions, Kluwer Academic Publishers.

Omid A.Sianaki, O Hussain, T Dillon, AR Tabesh - Intelligence, Modelling and Simulation (CIMSiM, 2010). Intelligent decision support system for including consumers' preferences in residential energy consumption in smart grid.

Power, D.J. (1996). What is a DSS? The On-Line Executive *Journal for Data-Intensive Decision Support* 1(3).

Power, D.J. (2000). Web-based and model-driven decision support systems: concepts and issues. in proceedings of the Americas Conference on Information Systems, Long Beach, California.

Power, D.J. (2002). Decision support systems: concepts and resources for managers. Westport, Conn., Quorum Books.

Reich, Yoram; Kapeliuk, Adi. Decision Support Systems., 2005, Vol. 41 Issue 1, p1-19, 19p.

Sauter, V.L. (1997). Decision support systems: an applied managerial approach. New York, John Wiley.

Silver, M. (1991). Systems that support decision makers: description and analysis. Chichester ; New York, Wiley.

Sprague, R. (1980). "A Framework for the Development of Decision Support Systems." MIS Quarterly. Vol. 4, No. 4, pp.1-25.

Sprague, R.H. and E.D. Carlson (1982). Building effective decision support systems. Englewood C41liffs, N.J., Prentice-Hall. ISBN 0-13-086215-0

Sprague, R.H. and H.J. Watson (1993). Decision support systems: putting theory into practice. Englewood Clifts, N.J., Prentice Hall.

Stanhope, P. (2002). Get in the Groove: building tools and peer-to-peer solutions with the Groove platform. New York, Hungry Minds.

Stephens, W. and Middleton, T. (2002). Why has the uptake of Decision Support Systems been so poor? In: Crop-soil simulation models in developing countries. 129-148 (Eds R.B. Matthews and William Stephens). Wallingford: CABI.

The Decision Support System for Agrotechnology Transfer

Wright, A; Sittig, D (2008). "A framework and model for evaluating clinical decision support architectures q". *Journal of Biomedical Informatics*. 41: 982–990. doi:10.1016/j.jbi.2008.03.009.

Zhang, S.X.; Babovic, V. (2011). "An evolutionary real options framework for the design and management of projects and systems with complex real options and exercising conditions". *Decision Support Systems*. 51 (1): 119–129.

❑❑❑

About the Author

Dr. Dileep Kumar born in 1980 in Jhunjhunu district of Rajasthan did his B.Sc. Agriculture (Hons.) and M.Sc. (Ag.) Extension from S.K.R.A.U. Bikaner (Rajasthan), India, in the year 2001 and 2004 respectively. His area of research during Masters Degree in the University was on Adoption of sprinkler system of irrigation in the Jhunjhunu District of Rajasthan. He pursued his Ph.D. degree from N.D.R.I. Karnal, (Haryana), on Critical Analysis of Dairy Co-operatives and Private Dairies in Rajasthan in 2012. Throughout his study, he secured I Position in all examinations and was awarded Merit Scholarships for under Graduate programmes, graduate and post graduate programmes. He is closely associated with many national and international professional societies of Extension Education and published many research and popular articles. He remained meritorious student during his academics. He is also pursuing PGJMC (Post Graduate in Journalism and Mass Communication) from IGNOU, New Delhi. Dr. Dileep Kumar started his career as scientist (Agricultural extension) in ICAR at ICAR RC for NEH Meghalaya in 2013. He is the Life Member of number of Professional Bodies of Academic Organisations and Institutions. He has authored ten books, and written more than 10 research paper published in Indian and foreign journals. At present author is principle investigator in a project entitled "Designing of participatory decision support system for horticultural crops in Meghalaya."He is Co-PI in various Institute and Externally Funded projects of social science. At present he is posted at ICAR-CAZRI,RRS-Jaisalmer and presently officer in charge (HOD) of the station. He has reviewed as many as 50 articles in different journals.

He has wrote 12 research paper as 1st author and 5 research paper as 2nd author. He has presented three papers in international conferences and 4 papers in national conferences. He got young scientist award in international conference organised by Astha society. He has authored 18 book chapters in different books. He has handled 5 projects, coordinated 5 trainings, authored 20 popular articles, delivered 30 lectures in different trainings as resource person. Edited 6 training manuals if different trainings.